FATE ON A FOLDED WING

The True Story of Pioneering Solo Pilot Joan Merriam Smith

TIFFANY ANN BROWN

A Lucky Bat Book

Fate on a Folded Wing:
The True Story of Pioneering Solo Pilot Joan Merriam Smith

Copyright ©2019 by Tiffany Brown

All rights reserved. No portion of this work may be reproduced or used in any manner whatsoever without the permission of the author. For permissions contact the author.

ISBN-13: 978-1-943588-82-4

Cover Design: Nuno Moreira
Published by Lucky Bat Books
10 9 8 7 6 5 4 3 2 1

Dedication

This book is dedicated to all of the world's adventurous,
brave, ambitious, and curious women, but most especially
to Joan Merriam Smith and Trixie Ann Schubert,
for whom this book is written.

= CONTENTS =

Introduction	vii
World Flight Itinerary	xix

PART ONE
1	A Flight that Changed History	3
2	The Backstory	18

PART TWO
	A Word From the Author	43
3	The Great Send-Off: From Oakland to Miami	46
4	Bearing Down on Brazil, Exploring Africa	57
5	From Asia and Indonesia, on to Australia	74
6	Investigating Amelia: Departing from Lae, Across the Pacific	101
7	The Fate of 51-Poppa	120

PART THREE
8	A Perplexing End	133
9	Justice for Joan	168

Epilogue	180
Acknowledgements	183

APPENDICES

A Letter from Trixie to Ruth Deerman,
   November 1, 1964   187
B Letter from Hillel Black to Trixie,
   February 17, 1965   189
C Trixie's timeline of correspondence between
   Joan and the NAA   190
D Letter from Roy A. Wykoff to *General Aviation
   News,* July 27, 1963   192
E Witness Account of Airplane Crash of Trixie-Ann
   Schubert and Joan Smith, February 21, 1965   194

About the Author   197

INTRODUCTION

Every now and again, you hear about someone who inadvertently discovers a priceless painting in their attic, or a valuable artifact lodged inside the wall of their home. In some cases, it's a treasure chest or a time capsule, offering a unique and mysterious glimpse into the past. In my story, I came across an unexpected relic: an old manuscript that had been sitting in a box for more than fifty years. Or I suppose you could say that the manuscript found me.

As a child, I often heard the story about my grandmother, Trixie, the foreign news correspondent who died suddenly in a plane crash with Joan Merriam Smith, the youngest woman to fly solo around the world. A few years ago, I couldn't have told you much about either my grandmother or Joan. There were no Wikipedia entries to speak of; a Google search would have produced next to nothing. The books Trixie had written were out of print, and the only pictures I could find of Joan were mostly buried in obscure historical archives. While I had access to old journals and photo albums, there was simply nothing in existence at that time in the form of audio or video. I wondered: if something didn't exist online, could it have really been that important?

As luck would have it, that question was soon answered for me in the form of blog post. While experimenting with WordPress in 2010, I decided to type up and share a letter that Trixie had written to her children before she died. I also included a link to the one story I could find about her plane crash, which came from an unexpected source: an issue of the Big Pines Volunteers newsletter from the town of Wrightwood, California.

Soon after publishing that post, I began receiving dozens of emails from people across the web. Perplexed and fascinated—but also excited—the frozen-in-time images I held in my mind of Trixie and Joan began to melt away once I realized that they were actual people who had made a very noticeable impact. Among the comments I received were the following:

"Your grandmother, Joan Merriam and the *Saturday Evening Post* article, 'I Flew Around the World Alone,' by Joan Merriam, that your grandmother helped Joan write, Aztec Aircraft at Long Beach Airport, and the other women who were your grandmother's contemporaries are all part of early memories for me. The women that your grandmother associated with, and was indeed one of, were a unique breed of women of adventure. It is ironic that the story of Joan's incredible life's ambition inspired by Amelia Earhart was accomplished at such a young age and then the important story never properly told."

"We knew Trixie—and your grandfather also. My former wife and I would often visit your grandparents at their home in L.A. We knew your mom because of my wife's involvement in the 99s, the International Organization of Women Pilots, of which your grandmother was a very active member. Had great convivial dinners at your grandmother's on a number of occasions, too. She was the kind of character one never forgets, filled with joy, intelligence, and good wit. She had quite an impressively uplifting sort of personality, always upbeat. Certainly a talented writer. She was a most memorable lady . . . ."

"Your grandmother and my mother were, as I understand it, friends back in the 50s and perhaps before. In 1948, my mother was diagnosed with breast cancer . . . the journey was documented by Trixie Ann Schubert and was written up in the

September 1956 issue of St. Joseph's Magazine.[1] My mother passed away in January 1957, two days after my sixth birthday. I have an original issue of that September 1956 St. Joseph's magazine that I have read and reread many, many times over the years. My father always had wonderful things to say about Trixie Ann, and I still remember the shock to him and, frankly, to all of us boys when we heard of her untimely death in 1965...."

"My mother also knew Trixie. They were very dear friends and I heard constantly about her in my life (mom is still alive at ninety-two and occasionally mentions her in conversation about the past). We leave our marks on this little planet for all time...."

Who knew that such a short and simple post could turn out to be such an unexpected force of connection?

When Joan and Trixie's plane crashed into the mountains outside of Los Angeles in February 1965, the incident made news headlines around the world. While it wasn't apparent at first (via Google), a quick search on Newspapers.com generated several hundred article mentions about the plane crash. You see, Joan and Trixie were not your ordinary 1950s housewives. Joan was not only the first person to fly solo around the world at the equator, but she had just completed a historic adventure retracing the equatorial Amelia Earhart route. Trixie was an author and foreign news correspondent who had recently returned from reporting behind the Iron Curtain in Russia while on assignment with the Associated Press. Both women were accomplished pilots at a time when females made up only 2.5 percent of all licensed private pilots: in 1960, there were only 3,425 female private pilots compared to 136,369 male pilots.[2]

---

[1] Based on my research, it appears this was a monthly magazine entitled *America's Catholic Family Monthly,* which was owned, edited, and published by Mount Angel Abbey in Saint Benedict, Oregon.

[2] "Pilot Certificate Statistics: 1960 – 2010," *Women of Aviation Worldwide Week*, accessed February 18, 2019, www.womenofaviationweek.org/wp-content/uploads/pdf/Pilot_Certificates_Statistics_1960-2010_table.pdf.

As it turns out, Trixie was working on a book about Joan's solo flight around the world, which was on track to be finalized during the very month in which they died. Ultimately, it was never published. While I had always known about this manuscript, I had never once thought to read it. It was around this time that my mother came across an original version of the manuscript, buried inside of a box within her garage, and passed a copy of it along to me. I decided it was time.

The Manuscript

The pages of Trixie's original manuscript, entitled *World Flight: Joan Merriam Smith*, are paper thin, emphasized with blue, typewritten ink, and covered with scribbled, handwritten notes. Few things about its condition would make any normal person want to jump in and read it, but for me the question of digging in was a no-brainer. With zero knowledge about the details of Joan's accomplishment, nor any understanding of Joan and Trixie's friendship, I wondered what I was going to learn. With a completely open mind, I was excited to find out.

However, once I started converting the text into digital format, I soon realized that I had a story with more questions than answers. Missing pages, unnumbered pages, out-of-order pages, hard-to-understand notes, and even a few missing chapters discouraged me, throwing me off my original plan to complete a straight publication of the story for historical purposes. Should I just forget about it and move on? How would I frame the book's original intent? What were the right questions to ask, and to whom could I ask them?

When I first pieced the story together, I instantly realized the challenge of attempting to construct a new story around an original one from over fifty years ago. But in their absence, the story's missing chapters and pages actually launched the journey. The initial disappointment of not having the full story is what actually led me to dig in and explore, ask the big questions, and uncover all sorts of new information that I would never otherwise have been able to find. If I'd had the full story, then I would never have been pushed to engage with it at this level . . . *and perhaps that was the point.*

While at first I was overwhelmed by the amount of details I did not know, the spirit of inquiry soon became the driving force that would carry me through—quite fitting for the granddaughter of a journalist. The first thing I did was to start having conversations with the very people who had initially reached out to me. Soon after, I was connecting with pilots, individuals familiar with Joan's story, and old friends of Joan's and Trixie's. To my surprise, I even located Joan's husband, then ninety-four-year-old Jack Smith, through a naval reunion video on YouTube. At the same time, I dove into the newspaper archives. I also pored over Trixie's volumes of journals, photo albums, and all of the materials she left behind related to the book. Ultimately, I catalogued a few hundred sources of information.

With each new read, lead, or conversation, additional insights were provided, creating more questions. Ultimately, the more questions I asked, the more evidence would present itself, and the more the story I thought I wanted to tell would change. This book is the result of that journey. Above all, however, this is Joan's never-before-told, first-person story about her around-the-world flight. After all, it was Trixie's primary goal to bring awareness to Joan's untold story.

Starting with the moment, in the manuscript, when Joan kicks off her global flight along the original Amelia Earhart route—from the same airport, on the same runway, on the very same day twenty-seven years after Amelia Earhart did it—those familiar with Earhart's flight will soon notice direct parallels to Earhart's own undertaking and realize that Joan's trip was always about more than just setting a record. From Long Beach to Oakland; across North America to Miami, down to South America; across the Atlantic Ocean to Dakar; through Africa, Pakistan, and India; to Indonesia, Australia, New Guinea, and across the Pacific; and finally to Oakland via Hawaii, Joan brings the reader with her into the cockpit, providing a front-row seat to her personal experience.

A Note on Historical Accuracy

Much has been written in recent years about the fact that our memories simply cannot be relied upon when it comes to recalling important events and information from the past. Often tainted with emotions, changing filters, and perceptions, memories are essentially reconstructed differently each time they are accessed based upon the moments in which we find ourselves at any given time. According to the *Journal of Neuroscience*, each time you recall an event, your brain distorts it: "A memory is not simply an image produced by time traveling back to the original event—it can be an image that is somewhat distorted because of the prior times you remembered it. Your memory of an event can grow less precise even to the point of being totally false with each retrieval."[3]

As such, historical relativism becomes a real problem when looking back on the events of the past. Historical relativism affects the way we interpret history by subjective factors characteristic of either the historian or the period in which the historian lives. It runs the risk of tainting the original story, overemphasizing less important aspects and/or leaving out some of the most important pieces altogether. Historians are quite aware of this issue, and many both teach and warn about the ways we can avoid reviewing history through colored lenses.

In my case, however, I had the good fortune of coming across a first-person account. And because I had literally zero background knowledge at the outset about the people, places, or things mentioned in this story, I came into this project with no preconceived notions. I had little problem challenging assumptions. Before too long, I was looking up terms like VHF, VFR, beacons, instruments, pistons, omni stations, and sextants; researching names like Goerner, Odlum, Dimity, Mantz, and Frey; locating places such as Khartoum, Bamako, Ouagadougou, Soekarno, and Surabaya. I wondered, what exactly had I gotten myself into?

---

[3] Donna J. Bridge and Ken A. Paller, "Neural Correlates of Reactivation and Retrieval-Induced Distortion," *Journal of Neuroscience*, Vol 32, Issue 35 (August 29, 2012): 12144-12151, https://doi.org/10.1523/JNEUROSCI.1378-12.2012.

The Lure of Aviation

According to one of the individuals who originally reached out to me through my blog post (a fellow pilot who had known Joan and has chosen to remain anonymous):

> This story is not only about Joan and Trixie, but all of the other great women of aviation, as well as aviation itself. From the first air-travelers—Marquis François Laurent Le Vieux d'Arlandes and Jean-François Pilâtre de Rozier—who flew in a balloon over Paris, France on November 11, 1783, to the Wright Brothers' first powered flight on the sands of Kitty Hawk, North Carolina, aviation would refuse to conform to earthly beliefs. Bessie Coleman, a woman of African American and Native American descent, became the first female certified pilot to lose her life in the pursuit of her dreams on April 11, 1926. Aviation itself keeps no records; knows no gender, no skin color, no politics, and, to date, no limits. It simply asks children of all ages to dream and come join it in a grand adventure... out there!

Among the most famous female aviators of all time, of course, is Amelia Earhart (or A.E., to which she is often referred in the aviation community). It may be hard to believe, given how tightly the persona of Amelia Earhart has been woven into the American cultural fabric, but while I had heard of Amelia Earhart and knew she was famous for attempting to fly around the world, I couldn't have told you any details about her world flight, why she did it, or what the specifics of her disappearance were prior to coming across Joan's story. In fact, I was pulled into this undertaking from a completely backwards—or I guess you could say "inside out"—angle, as I was exposed to Trixie's perspective first, then Joan's, and finally Amelia's. Learning about Trixie is what encouraged me to find out as much as I could about Joan, and Joan's lifelong interest in pursuing Amelia Earhart's goal is what prompted me to learn more about her.

I was soon surprised to learn that Amelia had flown around the world with a navigator (I assumed she had been in the plane by herself), that there had been a rough start to the original world flight (her plane made a ground loop and broke apart in Hawaii on the second leg of the original world flight), and that she had been married to a well-known publisher and author (George Putnam would go on to publish her book, *Last Flight*, following her disappearance). The details about Amelia Earhart's life percolated slowly into my own at first, but soon I was collecting so much information that I didn't know what to do with it!

Of course, I had always heard the theory that Amelia Earhart had either crashed into the ocean or died as a castaway on an island. But it wasn't until I read a letter that Trixie had saved about Amelia Earhart's disappearance that I heard anything about the Japanese capture theory. An original 1944 article from *The American Weekly* entitled "Amelia Earhart: How Long a Mystery?" piqued my interest, as did copies of old telegrams Trixie had saved about statements made by Amelia Earhart's mother. Among these items was a copy of WWII veteran and commissioned air force officer Paul Briand Jr.'s 1960 book *Daughter in the Sky*. A handwritten note from Briand inside the front cover reads, "For Trixie Schubert, with warm personal regards and best wishes for the Smith story and thank you for your nice visit at the Air Force Academy, Paul Briand." I paused. After learning that Briand's book was the first to contest the true nature of the Amelia Earhart disappearance, I wondered: why would Trixie have made a special visit to the Air Force Academy to visit with Briand in 1964?

After familiarizing myself with Trixie's manuscript, the materials related to the book, and her time spent behind the Iron Curtain, from there I would go on to read *Last Flight* to discover Amelia Earhart's own words for myself, along with many other books and articles about great female aviators in American history. After learning more about Beryl Markham, the famed racehorse trainer and African bush pilot who grew up in Kenya, for example, I next discovered the enigmatic Mary Petre Bruce, the record-setting 1920s speedboat racer, race-car driver, and pioneer around-the-world flier. I would also learn about lesser-known pilots like Fran Bera, a seven-time winner of the All Woman Transcontinental Air Race, and "Grace the Ace" Page, who performed comedy stunts for audiences and guests that included the likes of King Hussein of Jordan (also a pilot)[4]. It wasn't too long before I became completely enamored with each of these women's unique qualities, incredible contributions, and enigmatic, larger-than-life personalities. The more I learned about their collective journeys, the more my admiration and reverence deepened for each of them. Pretty soon, I was hooked on learning all I could.

Each of these women presents us with a unique set of important and timely lessons, reinforced by the personal choices she made about how to live her own life. Collectively, they allow for us to dream big, remind us that no challenge is too big to overcome, and show us that anything is possible. Their actions not only give us permission to trust our own inner voices, but encourage us to take control of the situations presented to us in life, even in the face of uncertainty and fear. And, above all, they push the limits of known boundaries in the spirit of adventure, progress, and desire. They are the patchwork of American greatness.

In the following pages, you will find a book that is divided into three parts. In Part One, I share the background of why Joan wanted to complete a world flight and how she prepared for it, and I try to shed some new light on an old controversy surrounding the sanctioning process to become the first woman to fly solo around the world (sanctioning is an important step for receiving official support, recognition, and verification when it comes to record-setting flights). I also cover Trixie's intentions for writing Joan's book and explore why it was never published in 1965. In Part Two, I share Joan's first-person account of flying around the world at the equator, the problems she faced, and the world as she encountered it. Finally, in Part Three, I explore the period following Joan's flight, the two untimely plane crashes she was involved in, and the unresolved questions that I was left with personally as a first-time reader of this story. I conclude with why this story matters and what we can learn from it. Beyond that, I've supplemented with facts, commentary, and memories to not only infuse it with other people's perspectives, but to help bring the story full circle.

---

[4] Mark Terry, "The Sky's the Limit: Remembering Grace Page," *Reflections Magazine,* Orange County, Florida, Regional History Center, Fall 2018, p. 17-19.

But, before we begin, I would be remiss not to share the very letter that kicked this whole thing off. (After all, coming across this letter back in 2010 intrigued me enough to post about it online, and, as a result, it became the catalyst for the entire experience of writing this book.) In advance of an impending solo flight from Kansas to California, in January 1964, Trixie wrote the following letter to her children, which was not to be opened until at least February 20 of that same year (per the note on the envelope that was saved with the letter). The letter would eventually be found and opened a year later, shortly following her death in February 1965:

Dear Monkeys Three—Patrice, Heidi, Norman:

I write this merely to emphasize what you already know—that every breath I breathe is with love for you three and with gratitude to God for having so blessed me with you.

I write, anticipating no problems on the solo flight I'm about to fly, but because I'm somewhat of a fatalist (that all happens by His permissive or positive will and that it is sufficient in life to fulfill our mission here on earth with the living prayer, "Let me be an instrument of Thy will" and nothing else matters) and consider all possible contingencies; as my Cub Scout (Norman) knows, wisdom lies in being prepared in all things as much as possible.

This flight is a challenge, one with which I feel capable of coping, or I would not make it. There is a selfish motive, too. I want to make you as proud of me as I am of you (and this is not a false pride). Our love must be bound by purpose, initiative, fulfillment, accomplishment, and you three are well launched on that track; we are not born in the image and likeness of God to vegetate.

You have nothing in life to fear, NOTHING, while you adhere to the magnificent faith bestowed on you in Baptism. Don't frustrate life by trying to understand all that you KNOW by feeling and experiencing. The mystery of faith implies that we accept some belief on faith alone; as we accept so many mundane matters because they "work" as we expect them to, even though we don't understand how (you know—electricity, growing grass, birth, love—yes, and pain and sorrow).

Temper the truths you hold in faith with tolerance for the beliefs of others, with integrity (don't imitate; be yourselves above all else and only then do you radiate charm, assurance, warmth, confidence).

As a child when my mother died, I felt life was over and it wasn't, though it was at the same time diminished and augmented. I felt it again when Daddy died when I was not yet adult and had hopes of his being with us as a family again. And when Nannie died, I wanted my world to end, never dreaming that she would come back to me in you three and that she never really left me. "I will be near, helping and praying for you whenever you need me," she promised, and she's there—helping, waiting.

Love binds—eternally, you shall discover if you haven't already. Obey Daddy and be as tolerant of his few foibles as he always has been of me. He is an exceptionally wonderful man, as if you didn't know. My love grows for him—always a good test of marriage. That's enough now of maternal commandments; no one yet has improved on the original ten anyway.

With you, and loving you—ALWAYS.

Mother

WORLD FLIGHT ITINERARY

B ELOW IS THE OFFICIAL LIST OF Joan's planned stops, taken verbatim from the promotional collateral related to the flight:

1. Tucson, AZ
2. New Orleans, LA
3. Miami, FL
4. San Juan, Puerto Rico
5. Paramaribo, Suriname
6. Natal, Brazil
7. Dakar, Senegal
8. Gao, Mali
9. Fort Lamy, Chad
10. El Fasher, Sudan
11. Khatoum, Ethiopia
12. Aden, Yemen
13. Karachi, Pakistan
14. Calcutta, India
15. Akyab, Burma
16. Rangoon, Burma
17. Bangkok, Thailand
18. Singapore, Malaya
19. Djakarta, Indonesia
20. Surabaja, Indonesia

21. Kupang, Timor
22. Darwin, Australia
23. Lae, New Guinea
24. Guam
25. Wake Island
26. Honolulu, HI
27. Oakland, CA

PART ONE

= CHAPTER 1 =

A FLIGHT THAT CHANGED HISTORY

On May 12, 1964, at the age of twenty-seven, Joan Merriam Smith became the first person in history to complete a solo flight around the equator, which marked both the fulfillment of a lifelong dream and a tribute to Amelia Earhart. Having obtained a pilot's license at the minimum possible age of seventeen, Joan was inspired to finish what Amelia Earhart had set out to accomplish in 1937, after having read a book at age fifteen about her attempt to circumnavigate the world.

Plagued with plane trouble from the start, Joan circled the globe in her beloved twin-engine Apache N3251P, which she affectionately referred to as 51-Poppa. Unlike Earhart, who had been accompanied by her navigator, Fred Noonan, Joan flew alone. Reaching all of Earhart's original flight stops (though passing over one rather than landing, due to landing restrictions), she even spent time in Papua New Guinea and Saipan to investigate the Amelia Earhart disappearance for herself.

After a series of setbacks due to mechanical troubles and persistent bad weather, some 27,750 miles and 56.5 days later, Joan crossed the finish line in Oakland, California. She became the first person to fly around the world solo at the equator and the first woman to fly a twin-engine aircraft around the world, among other accolades. By default, she became the youngest woman to fly around the world. She also set the record for the longest single solo flight around the world.

To put Joan's accomplishment into context, when it comes to other women who have completed solo flights around the globe, to date there have technically been just ten. And while there were at least two other women who completed solo flights during the Great Depression—before any woman had yet completed a solo flight across the Atlantic or Pacific Ocean—it wasn't until 1964 that an official solo flight around the world was made by a woman.

To provide a little history, the first known female to complete a circumnavigation of the globe in a plane was a British pilot by the name of Mary Petre Bruce. After setting many records in automobile racing, as well as a speed record for crossing the English Channel in a motorboat, taking up flying would become her next logical challenge. Not to be outdone by her race-car-driving husband Victor Bruce, who won the Monte Carlo Rally in 1926, in 1927, Mrs. Bruce entered the race. She traveled seventeen hundred miles for seventy-two hours without sleeping to finish sixth overall and place first among women. According to the book *Queen of Speed: The Racy Life of Mary Petre Bruce*, at the end of this race she was so overcome by exhaustion that she promptly fell asleep on top of the wheel. While it's hard to imagine the amount of perseverance, concentration, and dedication needed to push through such a feat, this experience would soon serve as great preparation for taking on a world flight.

After noticing a Blackburn Bluebird IV for sale in a window storefront while shopping in downtown London, Bruce decided to purchase it on a whim and *then* sign up for her first flying lesson, much to the surprise of others. After earning her pilot's license on July 26, 1930, she embarked on her world flight on September 25, 1930, from the Heston Aerodome.[5] Traveling through Eastern Europe, Syria, the Middle East, Thailand, China, and Japan, she would next have to fold up her plane's wings in order to ship it across the Pacific. Arriving in

---

[5] Nancy R. Wilson, *Queen of Speed: The Racy Life of Mary Petre Bruce*, (Chippenham, Wiltshire, England: Ex Libris Books, 2012), p. 84-97.

Vancouver, she would next fly down the West Coast of the United States to Los Angeles and across the country to New York—where she would be arrested for flying circles around the Empire State Building—before boarding yet another shipping vessel to complete her journey back to London. With minimal flying experience, Bruce remains among the least-experienced solo pilots to date in the category of world solo pilots, which makes her undertaking all the more astonishing. And her book really is that good!

In the same year that Bruce returned to London, twenty-four-year-old German pilot Elly Beinhorn set out to complete her trek around the world. Departing from Berlin in January 1931 in a Klemm KL-20 airplane, Beinhorn followed a western route. She made two equatorial crossings and faced several forced landings. She, too, shipped her plane across the oceans, returning to Berlin in June 1931. What made her flight all the more interesting is that she pulled this off while living in Nazi Germany. Because she married Bernd Rosemeyer, a famous race-car driver at the time, in 1935, the duo was exalted to celebrity status by the Nazi party despite the couple's wishes, as they were critical of the regime and feared for Germany's future. They stashed a portion of Rosemeyer's winnings into an American account in case they had to leave Germany to begin a new life.[6] (Interestingly, this fund would later become the basis for a commemorative Amelia Earhart stamp.) But, unfortunately, just six months later, Rosemeyer died in a fatal car accident, leaving Beinhorn behind to raise their ten-week-old son.

Following the accomplishments of Bruce and Beinhorn, on May 20, 1932, Amelia Earhart became the first woman to complete a solo crossing of the Atlantic Ocean. (Earlier, in 1928, she'd been credited with becoming first woman to fly across the Atlantic, but at that time she had only been a passenger.) Thirty-one years later, in 1963, Betty Miller became the first woman to fly solo across the Pacific Ocean. Once the trail had been blazed for women to travel by plane across the oceans

---

[6] Dr. Angelika Machinek, "Elly Beinhorn and the origin of the Amelia Earhart postage stamp," Ninety-Nines website, last modified June 1, 2009, https://www.ninety-nines.org/elly-beinhorn.htm.

alone, the stage was set for a female to complete a full circumnavigation of the globe. But who would be next?

As it turns out, there was at least one woman already working to answer this question, and her name was Dianna Converse Cyrus Bixby. Having attempted a world flight on three different occasions beginning in the year 1949, Bixby was ultimately set back due to a combination of mechanical issues, bad weather, and inability to raise capital. It was during this time that Bixby and her husband started Bixby Airborne Products, a commercial charter service specializing in the transport of perishable items. While working to raise money for a fourth flight after her third failed attempt in 1954, tragedy struck. While flying between Long Beach and Mexico, Bixby flew into a violent storm. Low on visibility and low on fuel, she attempted to land, but her plane crashed into the ocean. The wreckage was found the following day in only fifteen feet of water, with Bixby still strapped into the cockpit. Even though she was not able to realize her dream, her efforts were not lost; in fact, they would lay the groundwork for future pilots.[7]

Enter Joan Merriam Smith. Immediately preceding Bixby's accident in November 1953, Joan had just wrapped up high school, earning her pilot's license with the ultimate dream of becoming the first person to complete the Amelia Earhart route. By January 1954, she joined the Miami chapter of the Ninety-Nines. A few years later, in Florida in 1958, she met Jack Smith, a U.S. Navy officer and pilot. Joan moved to the West Coast, and the two were married in 1960. Coincidentally, Jack had worked for the Bixbys. Despite knowing all too well the risks, he supported Joan in her endeavor to proceed with completing her personal ambition of a world flight.

By 1964, Joan had completed her flight around the world, as had American Jerrie Mock, who followed a shorter route across the northern hemisphere. While Joan received accolades for becoming the person to fly solo around the world at the equator, Mock finished ahead of Joan, claiming "first woman to complete a flight around the world solo" honors

---

[7] Claudine Burnett, *Soaring Skyward: A History of Aviation in and Around Long Beach, California* (Bloomington, IN: AuthorHouse, 10/27/11), p. 168-171.

(much more on this in Chapter 2). British pilot Sheila Scott, an actress who had taken flying lessons on a dare, would next go on to complete three separate flights around the world in 1966, 1969, and 1971. Setting abundant records, Scott became the first solo flier to cross the North Pole; she also set a new record for the longest solo flight ever made.[8] In 1980, British pilot Judith Chisholm set a record for becoming the fastest woman to fly solo around the world. And in 1989, Gaby Kennard became the first Australian woman to achieve the feat.

Moving on to the 21st century, American pilot Jennifer Murray became the first woman to solo the world—this time in a helicopter—in the year 2000. British pilot Polly Vacher next flew around the world in 2001, and again in 2004, to raise money for the charity Flying Scholarships for the Disabled.[9] In 2003, and again in 2011-12, CarolAnn Garratt completed solo flights to raise awareness for amytrophic lateral sclerosis (ALS). In 2016, Julie Wang became the first Asian woman and first Chinese person to pilot an aircraft around the world solo.[10] And most recently, in 2017, Afghan-American pilot Shaesta Waiz became the youngest woman to fly around the world in a single-engine plane with the goal of inspiring the next generation of STEM (science, technology, engineering, and mathematics) and aviation professionals.[11]

---

[8] "Sheila Scott; Actress Learned to Fly and Set Many Records," Associated Press, *Los Angeles Times*, October 22, 1988, accessed 2/18/19, http://articles.latimes.com/1988-10-22/news/mn-11_1_sheila-scott.

[9] Polly Vacher, *Wings Around the World: the exhilarating story of one woman's epic flight from the North Pole to Antarctica* (London: Grub Street: 2006), p. 7.

[10] "Award-Winning Documentary on Silver Airways Pilot Julie Wang to be Featured at Fort Lauderdale International Film Festival and Chinese American Film Festival in Los Angeles," Yahoo Finance, last modified October 24, 2018, https://finance.yahoo.com/news/award-winning-documentary-silver-airways-133200971.html.

[11] Diane Tedeschi, "Around the World in 145 Days," Smithsonian Air and Space Magazine, June 2018, https://www.airspacemag.com/as-interview/interview-180968989/.

The Ambitions of Earthrounders

Before I encountered Joan's story, I wondered, why would someone would ever want to complete a solo flight around the world? How much does it cost? How do you raise the money, obtain the clearances, or even get started? How much experience do you need? As it turns out, preparing for a world flight is no simple task.

According to Australian pilot Claude Meunier—who completed his own solo flight around the world in 1996 and who maintains Earthrounders, the international association for pilots who have completed flights around the globe—it is impossible to know about every solo flight that has been made because many are so difficult to find. There is no official register of those who have flown around the world. But of the 132 world solo flights he's tracked, he does give special credit to those who completed flights before the introduction of the Global Positioning System (GPS) in the 1970s. In describing how GPS affected aviation, he writes:

> For a few hundred dollars and without the need for any deep knowledge, one can know their position within a few feet anywhere on earth by day or night. Today, without diminishing the merit of pilots flying on long flights, or flying around the world, their task is now less difficult than that experienced by the pioneers of the 50's and the 60's. [. . .] There are many documented "around-the-world" flights which have been made by military aircraft and airlines, but these all had enormous support both in the air and on the ground, and were, in the main, huge publicity campaigns, costing millions. My interest lay in the traditional solo flights undertaken by the pioneers, their problems, how they were solved, and of their success in succeeding alone."[12]

Based on his own experience of flying around the world alone, combined with what he's learned from his fellow earthrounders, Meunier

---

[12] Claude Meunier, "Introduction," Soloflights.org, September 24, 2013, http://www.soloflights.org/intro_e.html.

offers some great advice for those looking to get started on planning a flight around the world. Among the many considerations, a pilot must, first and foremost, have the proper training and licenses, the right mix of experience, and the courage to face the unknown. Scores of money must be raised, a support team identified, flight stops charted, clearances granted, visas and passports obtained. When choosing a plane, a pilot has to decide if he or she wants a single or double engine—ideally the engine should be overhauled—and a proper radio must be installed. Additional fuel tanks may also need to be added as there are many fuel, speed, and distance calculations to be made. Special attention during the planning process should be given to weather, airspace restrictions, and areas around the globe where travel is not permitted. The plane should also be fully stocked with survival gear, including a life jacket; life raft; space blanket; marine radio; thermal suit (if necessary); and emergency kit with water, food, flares, and more.

Solo pilots embarking on these flights also have to be ready for anything and willing to forego everyday comforts for long periods of time. What should they eat or drink? What happens if they get sick, need to use the restroom, become too tired, or get too hot or too cold? What will the plan of action be if their radio stops working, if they run out of gas, or face sudden mechanical troubles in the air? Is the purpose of the flight to set a record, raise awareness for a cause, learn more about a particular part of the world, gain recognition, or overcome a personal challenge?

In terms of the process one should follow when attempting to set a new record, Meunier explains that records in aviation are verified and kept by the Fédération Aéronautique Internationale (FAI). To start, one must first check to make sure that the particular record has not already been set. The next step is to contact the National Aero Club for a starter kit and then apply for a record permit. Once that permit is issued, no one else can apply for a set period of time. He continues: "Apart from the permit application, it is best not to broadcast one's intention of setting or breaking a record. Competitors may otherwise get in first."[13] On his

---

[13] Ibid.

website, he cites the example of Australian Dick Smith, who publicly announced his intention to be the first to fly around the world in a helicopter in 1982. Upon reading an article about the interview, Ross Perot Jr. (son of the former presidential candidate Ross Perot) decided that such a record should go to the USA and subsequently flew around the world before Smith was even ready.

Taking all of this into consideration, it's not surprising that the question Meunier most often receives is, why would anyone even want to complete a flight around the world at all? After all, it's expensive, risky, and time-consuming, and there are many unknowns. He explains:

> I have been asked many times, why fly around the world alone in a light aircraft? There is no easy answer. Why do mountaineers want to climb Mount Everest? Because it is there, because it is the highest, because it is their ultimate challenge. I think, it is the same with navigators. After going further and further, they seek the ultimate: a journey around the world. Whether it is sailing a boat or flying an airplane, the challenge is the same. Who is the navigator who never had this dream?[14]

In short, the pilots who choose to embark on such undertakings do it simply because they have to. It's a calling, an adventure, a one-of-a-kind learning experience, and a personal challenge all wrapped into one.

The Book That Never Materialized

Take a moment to imagine a time before handheld calculators, cell phones, email, GPS, the Internet, jumbo jets, personal computers, the Stealth Bomber, video games, and virtual reality. For those born after 1965, that's hard to do; this was what life was like before 1965.

The 1960s were dominated by the Civil Rights Movement, the Cold War, the Space Race, the Cuban Missile Crisis, the Vietnam War, and the Beatles. In 1960, a U.S. spy plane was shot down in Russia, leading

---

[14] Ibid.

to the Francis Gary Powers trial. In 1961, President John F. Kennedy Jr. warned citizens to prepare for a possible nuclear attack, and by 1965, the Soviets had tested a fifty-megaton hydrogen bomb, resulting in the biggest explosion in history. In 1963, Martin Luther King Jr. gave his famous "I have a dream" speech, and Kennedy was assassinated. Also in 1963, Betty Friedan launched the modern feminist movement with her critique of the role of women in society, and Gloria Steinem went undercover as a Playboy Bunny to write an infamous expose of about the harassment of and injustices against women. The Sixties also introduced "the pill," the Super Bowl, and *Rolling Stone* magazine to American society. And in 1969, Neil Armstrong became the first man to walk on the moon. A new era was emerging, and Joan and Trixie were on the leading edge of progress.

Trixie loved to write, and she wanted to write Joan's book in 1964. Trixie was more than just an author: she was a risk-taker, knowledge-seeker, and adventurer. She became a pilot in 1944 and raced in both international and transcontinental air races. Prior to working on Joan's book, Trixie authored three books, including *A Bell in the Heart*, which merited a foreword by Admiral Arleigh Burke, chief of naval operations under the Eisenhower administration. Having worked as a freelance foreign news correspondent in America, Europe, Asia, and Africa, she had a varied career in writing as the editor of a weekly newspaper; a radio announcer; a news writer for *The Milwaukee Journal* and AM, FM, and TV stations; and an aviation columnist.

In advance of traveling behind the Iron Curtain during the height of the Cold War in 1960, she learned to speak German and Russian in order to accept an Associated Press assignment as part of an experiment in foreign news reporting. In addition to her time working and traveling as a journalist, she married Dr. Delwyn Schubert, who coordinated education programs for the U.S. Air Force in Africa, Asia, and Europe. Together they had three children. In the late 1950s, they were stationed at Wiesbaden Air Base in Germany.

Drawing from pieced-together sentiments from Trixie's personal journals, it's apparent that she was eager and excited to write Joan's book. Following Joan's return from her around-the-world flight, on

May 17, 1964, Trixie wrote: "In evening I drove with Shirley Thom to Redondo Beach's 'Plush Horse'[15] for homecoming banquet for Joan Merriam Smith. Was seated at head table near Joan. I would most like to get a crack at writing her book."[16]

Then, on September 21, 1964, she wrote: "Joan Merriam Smith here. Gradually I'm building up enough data to get her book underway."[17]

On October 14, 1964: "With Joan Merriam and her mother all day in Long Beach."[18]

A few months later, on November 4, 1964, she noted: "Finally sent off the manuscript of Joan's flight. Worked all day yesterday with Joan at Long Beach. Jack took us to lunch."[19]

Ah... Jack. I first learned about Joan's husband, Jack, in a YouTube video from his naval reunion coordinator in which Jack reflected on his experience of having served as a commanding officer on the minesweeper *USS Endurance*. At age ninety-four, Jack—Lieutenant Commander Captain Marvin "Jack" Smith Jr., as he is known—shared sentiments about Joan and Trixie, his dreams as a young man, and the pain he experienced following the tragedy of losing his wife and friend. But, for now, let's get back to the story about the unpublished book.[20]

Upon sending off her manuscript, *World Flight: Joan Merriam Smith*, to the publisher in November 1964, Trixie wrote a letter to Ruth Deerman (see Appendix A), the international president of the Ninety-Nines. Originally founded in 1929 with just ninety-nine members, the Ninety-Nines of today is an organization made of up licensed female

---

[15] The Plush Horse Inn opened at 1700 Pacific Coast Highway in Redondo Beach on Aug. 3, 1960. The facility near the corner of PCH and Palos Verdes Boulevard had two adjacent eateries, a larger restaurant known as the Plush Horse, and the Plush Pony coffee shop.

[16] Trixie Schubert's personal journal, May 17, 1964.

[17] Ibid, Sept. 21, 1964 entry.

[18] Ibid, Oct. 14, 1964 entry.

[19] Ibid, Nov. 4, 1964 entry.

[20] "Global Flier, Husband Reunited After 7 Months," *Los Angeles Times*, July 29, 1964, p. 20.

pilots from forty-four countries. Amelia Earhart served as the organization's first president. The letter reads:

Dear Ruth,

Thank you for your inquiry regarding the book on Joan's flight. I completed it last night, but it required some literal night and day writing. I have a publisher who is most interested, and Lowell Thomas has agreed to write a foreword for the book. It is called: "World Flight: Joan Merriam Smith" with the subtitle "Amelia Earhart route." Publishers have been known to change the titles so we only assume this will be the published title.

Trixie-Ann Schubert[21]

Born in 1892, Lowell Thomas was not only a friend of Trixie's, but a longtime broadcaster for CBS radio and the author of more than fifty books, including *With Lawrence in Arabia*. He launched Cinerama, the first widescreen motion picture technique that revolutionized the motion picture industry. He was also widely known as a globe-trotter, among the first to travel in and write about Afghanistan and other countries. He died in 1981. His son, Lowell Thomas Jr., was an author, adventurer, glacier pilot, filmmaker, and, in the early 1970s, an Alaskan state senator. He was also the third lieutenant governor of Alaska.[22] Because no known copy of Thomas' foreword to Trixie's book exists, one can only speculate how his foreword might have read.

A few months later, Trixie was making good progress on her book, having rewritten a few chapters at the request of her publisher. Since she had first sent off the manuscript, Joan was involved in a plane crash, so Trixie added a chapter to the book about this incident. A letter to her publisher, Ward Ritchie Press, dated January 19, 1965, reads:

---

[21] Letter from Trixie Schubert to Ruth Deerman, November 1, 1964.

[22] "Thomas, Lowell," The National Aviation Hall of Fame, accessed Aug. 1, 2018, https://www.nationalaviation.org/our-enshrinees/thomas-lowell/.

> Mr. Jonathan Ritchie,
>
> Here are a few press pictures of the hundreds on the world flight available for publisher selection. And, when you wish, Joan will meet here at the house some evening for a showing of the color slides. In the meantime... here is the final chapter (Last Flight for Apache 51-Poppa)[23] and some rewritten chapters of the original draft. Shall have the chapters that need rewriting finished and delivered to you soon. But if you've an opportunity to read the final chapter on the loss of the plane you might wish to comment on the style and content, or suggested improvement.
>
> Trixie-Ann Schubert[24]

Approximately two weeks later, Trixie noted that the *Saturday Evening Post* was interested in publishing the "Last Flight" chapter from her book. A few months prior, the *Saturday Evening Post* had run a feature story about Joan's around-the-world flight in its July 25-August 1, 1964 issue entitled "I Flew Around the World Alone." The "Last Flight" article was likely a follow-up story to that piece, presumably being pitched to help tease out the forthcoming book. In her journal, she wrote:

> February 3, 1965—Saturday Evening Post (Hillel Black) wrote that they would be interested in my final chapter on Joan Merriam. Jonathan Ritchie from Ward Ritchie Press called to say the book committee is enthusiastic about my book and I await only Monday's discussion with their distributors before signing a contract with me [sic].[25]

---

[23] Note: a form of this final chapter was later published in a March 1966 issue of Aircraft Owners and Pilots Association magazine, *The AOPA Pilot*, entitled "The Fiery End of 51 Pops" under Joan's byline.

[24] Letter from Trixie Schubert to Ward Ritchie Press, January 19, 1965.

[25] Trixie Schubert's personal journal, February 3, 1965.

But just two weeks following this journal entry, Joan and Trixie died. Whatever the outcome of the discussion was with the book distributors on February 8, we will never know. In a strange, cruel coincidence, on the *very day* of Joan and Trixie's fateful plane crash, the *Saturday Evening Post* sent Trixie the following rejection letter (see Appendix B):

February 17, 1965

Dear Mrs. Schubert:

Thank you for allowing us to consider the last chapter of your book which describes the death of Joan's airplane. I thought it was well done. However, after some thought and consideration here I have come to the conclusion that it would not work for us. But we certainly appreciate your thinking of the Post.

Cordially,
Hillel Black, Senior Editor[26]

If the usage of the word *death* jumped out to you in reading this response from Mr. Black, you're not alone; it did the same for me. Needless to say, without a champion or a spokesperson to presumably oversee its publication, the book about Joan's around-the-world flight never saw the light of day.

An Untimely Fate

Just seven months after the completion of Joan's pioneering flight, on January 9, 1965, she narrowly survived a plane crash in the California desert that destroyed her historic plane. And just a few weeks later, on February 17, 1965—when she was well on her way to becoming a legend—Joan died. The details of her fatal plane crash are outlined in this February 18, 1965, *Associated Press* article:

---

[26] Letter from Hillel Black to Trixie Schubert, February 17, 1965.

The first woman to fly the equatorial route around the world is believed to have piloted a small plane that crashed in the San Gabriel Mountains Wednesday, killing the two women aboard. Although the coroner's office declined official identification until her husband views the badly burned bodies, the husband said he had no doubt his wife, Joan Merriam Smith, 28, is dead.

Authorities believe the other woman was Trixie Ann Schubert, 43, of Los Angeles, who was writing Joan's life story. Sheriff's deputies and federal authorities reported that the women took off toward the area of the crash from nearby Long Beach.

Joan and Jerrie Mock of Columbus, Ohio, caught the world's imagination last year with their race to become the first woman to circle the globe. Mrs. Mock finished her 'round the world trip before Joan, but Joan claimed victory because she flew the so-called Amelia Earhart or equatorial route, which was 4,000 miles farther. [. . .]

Joan just recently walked away from a crash landing in the California desert. That crash destroyed the light plane she used on her epic flight, a twin-engine Piper Apache. Wednesday she was flying a Cessna 182. "After all those years and all those thousands of hours, why did it have to happen like this," said her mother, Ann Merriam, of Miami, Fla."[27]

---

[27] "Flier Joan Smith, Writer Are Killed." Associated Press, *The Times* (Munster, Indiana), Feb. 18, 1965, p. 2.

The news of Joan's and Trixie's deaths sent shock waves through the aviation community, and at that time, this is where their story ended. However, perhaps the tale of Joan and Trixie was not meant to be told *then*. What began as a small curiosity on my part soon turned into a persistent and nagging need to know more. With a copy of an original, typed manuscript and a box full of pictures, letters, newspaper articles, telegrams, and clues as to where this story was headed, it was almost as if this story was just waiting to be told.

= CHAPTER 2 =

THE BACKSTORY

"The fact is that the career of one who indulges in any kind of flying off the beaten path is often complicated. For instance, if one gives out plans beforehand, one is likely to be charged with publicity seeking by those who do not know how difficult it is to escape the competent gentlemen of the press. On the other hand if one slips away as I have generally tried to do, the slipper away invites cat calls from those who earn their living writing and taking photographs. So I am hoping the pros and cons of the whole undertaking can wait until it is finally over. If I am successful, the merits and demerits can be threshed out then. If not, someone else will do what I have attempted and I'll pass the problem onto him, or her."

— Amelia Earhart[28]

IN 1964, JOAN'S GLORY OF COMPLETING the world flight was dimmed because the first woman to achieve substantial and international press for finishing a flight around the world solo went to Jerrie Mock, an American housewife and recreational pilot. As fate would have it, both Joan and Jerrie would take off for their world flights within just two days of one another. But how did two such carefully planned flights, and of such magnitude, come together only two days apart? To arrive

---

[28] Amelia Earhart, *Last Flight* (New York, NY: Harcourt, Brace & Co., 1937), p. 58-59.

at that answer, let's begin by taking a look at the known intentions, the observable events, and outcomes.

Jerrie Mock's Path to Flight

According to research completed by Amy Saunders, a former reporter for *The Columbus Dispatch*, Jerrie Mock's decision to fly solo around the world began with a conversation one night at the dinner table. As the story goes, Jerrie complained to her husband, Russ, that she was bored with staying at home all day (as a mother with three children). He suggested, "Maybe you should get in your plane and just fly around the world." This would soon turn out to be a challenge that she willingly accepted.

In a 2014 article entitled "How an Ohio Housewife Flew Around The World, Made History, And Was Then Forgotten," Saunders provides a summary of this undertaking:

```
With her family's encouragement, Mock called
the National Aeronautic Association, just hop-
ing to obtain maps for an around-the-world trip,
and learned no other woman had completed the
flight. [. . .] Mock devoted more than a year
to preparation, meeting with Air Force officers
who secretly thought she was crazy but helped
her chart 19 stops. To acquire permission for
landings, she sorted through regulations and con-
tacted foreign embassies to ensure she wouldn't
be mistaken for a spy.29
```

The article goes on to talk about how Jerrie's husband, an advertising and public relations executive, had secured the *Columbus Dispatch* as

---

[29] Amy Saunders, "How An Ohio Housewife Flew Around The World, Made History, And Was Then Forgotten," *Buzzfeed*, April 12, 2014, https://www.buzzfeed.com/amyksaunders/the-untold-story-of-the-first-woman-to-fly-around-the-world.

a sponsor to pay $10,000 toward the flight (the equivalent of nearly $82,000 in 2019 dollars per the Bureau of Labor Statistics' CPI Inflation Calculator) in exchange for exclusive news about "the flying housewife." This strategy was not uncommon at the time; in fact, Amelia Earhart's own husband, George Putnam, had doubled as her chief promoter. As one of the most successful promoters of the 1930s, Putnam knew a thing or two about public relations and how to help gain exposure for Amelia in the mainstream. In addition to being a WWI veteran and seasoned international explorer, he was also a Harvard-educated newspaper editor, a twice-elected town mayor, and the grandson of a prominent publisher.

In 1927, Putnam published the best-selling book *We*, about Charles Lindberg's historic nonstop, transatlantic solo flight from New York to Paris. Shortly thereafter, he was contacted by Amy Guest, a wealthy American living in London who wanted to sponsor the first-ever flight by a woman across the Atlantic Ocean. Putnam suggested Amelia and ultimately connected her to the opportunity. Following the successful format of Lindbergh's book, Putnam went on to publish Amelia's account of becoming the first woman to fly across the Atlantic Ocean.[30] That book, entitled *20 Hrs., 40 Min.: Our Flight in the Friendship*, was published in 1928.[31]

As with Putnam, Russ worked behind the scenes to secure pivotal sponsorships for Jerrie, with ambitions for turning Jerrie into a household name. In a December 9, 1963 letter to a friend named Ralph about the progress being made on Jerrie's world flight preparations, Russ talked about the process of securing sponsorships and his goal of getting Jerrie into the spotlight. He wrote:

---

[30] In 1928, Amelia Earhart was the first woman passenger to complete a flight across the Atlantic, but she was accompanied by two co-pilots. It wasn't until May 20, 1932, that Amelia Earhart became the first woman to complete a solo crossing of the Atlantic Ocean.

[31] "Putnam, George Palmer (1887-1950)." Purdue University Libraries, Archives and Special Collections, accessed September 2018, https://archives.lib.purdue.edu/agents/people/1288.

Gosh, I feel good. Putting this together has been a real project . . . a real challenge. My background in advertising and public relations has been the big factor for success because I could talk to these people on a sensible basis [. . .]. If we could get her around the world, then keep her name hot long enough to knock off a second record, we'll be on the way to having a good commercial property.[32]

In the letter, he also talks about being worried that someone else (like Betty Miller) might jump into her plane and take off before his team is ready to get Jerrie into the air. He mentions checking in with the NAA to see what they knew.

As luck would have it, shortly after securing their sponsorships, Jerrie's team would learn about Joan. In an October 2014 article in *The Washington Post* about the life and legacy of Jerrie Mock—who passed away in 2010 at the age of 88—a brief mention of Joan appears: "When she learned that another female pilot, Joan Merriam Smith, was attempting a similar flight, Ms. Mock moved up the date of her departure by two weeks. Smith took off from Oakland, Calif., on March 17, two days before Ms. Mock."[33] Feeling the pressure of what had become a race, Jerrie scrambled to depart for her world flight earlier than planned. The fear was that if Joan gained a considerable lead, *The Columbus Dispatch* might pull its sponsorship for Jerrie, effectively canceling the flight.

Joan's Commitment to a Dream

On November 22, 1963—the day President John F. Kennedy was shot—Joan decided to turn her dream into a reality and put her life savings of $10,000 down to purchase the plane she planned to fly around the world. At age twenty-seven, Joan was determined to follow her dream

---

[32] Typewritten copy of a letter from Russell C. Mock to a friend named Ralph about Jerrie's world flight preparations, December 9, 1963, courtesy Taylor Phillips.

[33] Matt Schudel, "Jerrie Mock, first female pilot to fly solo around the world, dies at 88," *The Washington Post*, October 1, 2014, https://www.washingtonpost.com/national/jerrie-mock-first-female-pilot-to-fly-solo-around-the-world-dies-at-88/2014/10/01/2b33543a-4984-11e4-a046-120a8a855cca_story.html.

and make history happen. In an August 1964 Ninety-Nines chapter newsletter, she reflects:

> This dream, not a brainstorm, began at the age of 16 and never let go completely. Learning to fly at 16, getting my private license at 17, commercial, instrument, instructions rating at 18, and finally the ATR at 23 to become the first gal to receive that rating at the minimum age of 23. Dedication to aviation from high school days, long hours at airports, jobs varied from instructing flying 1954-1958, corporation pilot 1958-1962, charter and test flying at present. But the one long desire never left me: the urge to fly around the world at the equator, and follow Amelia Earhart route. I had admired her courage and determination since I read her book, Last Flight at age 15. Silently, the thought of some day making that same trip came into mind, and into reality in 1964 after 11 long years of wishing and trying.[34]

In the eleven years leading up to her world flight, Joan was very focused on preparing for one. Following the death of her father, in 1952, Joan had her first experience inside of a plane on a flight between Detroit and Miami. While at first terrified, she felt the experience was life changing. After talking with the pilots, she decided that taking flying lessons at twelve dollars an hour would be more worthwhile than continuing with baton-twirling lessons at six-fifty per half hour.[35]

---

[34] Joan Merriam Smith, "Tribute to a Star: Flying the A.E. Route," *Ninety-Nine News*, August/September 1964, p. 12, https://www.ninety-nines.org/pdf/newsmagazine/19640809.pdf.

[35] Joan Merriam Smith, "I Flew Around the World Alone," *Saturday Evening Post*, July 25, 1965, p. 78. https://s3.amazonaws.com/files.saturdayeveningpost.com/uploads/reprints/I_Flew_Around_the_World_Alone/index.html.

By the age of seventeen, Joan had already obtained her private pilot's license and soloed for thirty-five hours.[36] At seventeen, Joan was also the youngest entrant in the All Woman's International Air Race. By age twenty-three, she became the first woman to achieve the Airline Transport Rating (ATR) at the minimum possible age. (It is worth noting that because women could not yet pilot planes commercially at that time, acquiring the ATR was more about proving to herself that she could do it than anything else.) And by age twenty-seven, after plans had fallen through twice, she finally secured the plane for her venture.

In the months leading up her to world flight, Joan spent numerous hours poring over her flight charts, which lay stretched out all across the living room floor of her house. With twenty-eight legs, each approximately one thousand miles, the entire trip's flight charts, when pieced together, were eighteen feet long. Later, she would recall, "I had a 12-foot strip of aluminum floor molding which I used as a straightedge, and I spent about six hours figuring each 1,000 miles—more than 150 hours of plotting in all—marking off checkpoint, double-checking radio frequencies, looking for errors."[37] But approximately two months prior to takeoff, Joan would also soon learn that she wasn't the only woman seeking a record to become the first female to fly solo around the world.

The Coveted NAA Sanction

Now that we've established how both Joan and Jerrie got to this point in their flying careers, two questions present themselves: when did each woman formally learn of the other's ambition to fly solo around the world, and how did the process of obtaining a sanction actually unfold? For individuals wishing to set new flight records back in 1964, the first step in the process was to make sure that the particular record had not already been set. If all was clear, then applying for a sanction

---

[36] "Application for Membership for Joan Merriam," The Ninety-Nines Inc. International Organization of World Pilots, January 28, 1954, courtesy 99s Museum of Women Pilots.

[37] Joan Merriam Smith, "I Flew Around the World Alone," p. 80.

through the National Aeronautic Association (NAA) came next. As it pertains to flight records, being awarded a sanction was—and still is—important because it not only legitimizes the undertaking, but it guarantees exclusive rights to the individual attempting a particular record within a specified period of time.

In the case of Joan and Jerrie, the NAA had never before run into the problem of two females wanting to pursue around-the-world flights at the same time. So how did the NAA handle it? While the details leading up to the granting of the sanction are fuzzy, both Joan and Jerrie generally have their own interpretations of events. Are their accounts simply two different sides of the same story, or are they two different stories altogether? Ultimately, you will have to be the judge.

According to the book *Racing to Greet the Sun: Jerrie Mock and Joan Merriam Smith Duel to Become the First Woman to Fly Solo Around the World*, Jerrie maintains that she had been planning for her around-the-world flight since at least 1963. From documents shared by Taylor Phillips—the book's author, who also personally interviewed Jerrie before she passed away—this is true. In January 1963, NAA representative MJ (Randy) Randleman wrote a letter to Russ Mock clarifying what would be necessary for his wife to set an aviation record for "Speed Around the World."[38] Included with this letter was an "Application for Sanction of Record Attempt" and a "Schedule of Fees and Charges," among other enclosures.

Joan's account as it pertains to her correspondence with the NAA is a bit different. According to a timeline catalogued by Trixie that was included as part of the manuscript (see Appendix C), Joan initially wrote the NAA for information on August 4, 1963, to inquire about the process. By mid-September 1963, she had asked what was necessary to have a world flight sanctioned, and in October 1963, she was asked to submit a route itinerary. She mailed a complete itinerary on December 17, 1963, expecting sanction forms that she should fill out and send by return mail. In January 1964, Joan flew to Miami to help her mother, who had

---

[38] Letter to Russell C. Mock from M.J. Randleman at National Aeronautic Association, January 8, 1963, courtesy of Taylor Phillips.

fractured her skull and needed brain surgery. Around this time, she became concerned that there were still no sanction papers.

Approximately one full year after Jerrie had initially inquired with the NAA, she received an unexpected call from Randleman in Washington, DC, to let her know that there was another woman in the running to achieve the same sanction. Taking no chances, the Mocks caught the next commercial flight to Washington that evening, and by nine o'clock the next morning, Jerrie was waiting for the NAA office to open. There, she got all of her papers filled out and stamped.[39] A copy of Jerrie's original NAA Sanction Application shows that the form was signed on January 6 but that it wasn't formally approved by Randleman until the tenth.[40] Did Jerrie sign it in Ohio and hand carry the papers to Washington, DC, or did she fly to Washington with blank forms and *then* sign them in person? Did the call from Randleman prompt her trip, or was she already planning to be in Washington? Without knowing for certain the exact date of her flight, it's hard to say, though a copy of some handwritten notes by Jerrie, which Phillips shared, may provide some clues. The note for January 6, 1964, reads as follows: "NAA applies for passport, Jerrie in D.C., receives sanctions."[41]

As it turns out, during the same week that Jerrie was in Washington, DC, Joan would for the first time learn of Jerrie's impending flight. While shopping for equipment for her around-the-world flight at Pan Tronics in Florida, she was tipped off after a brief conversation with a local radio tech. From the document labeled "Joan's Timeline with NAA" come the following details (taken verbatim):

> January 7—Jack Riley at Fort Lauderdale. Was looking for a high frequency transceiver. Bob Urico of Sun Aire Electronics

---

[39] Taylor Phillips, *Racing to Greet the Sun: Jerrie Mock and Joan Merriam Smith Duel to Become the First Woman to Fly Solo Around the World*, published by the author, Kindle Version (January 19, 2015), chapter 1.

[40] NAA Application for Sanction, Jerrie Mock.

[41] A copy of a page including handwritten notes by Jerrie Mock about her world flight, courtesy of Taylor Phillips.

at Fort Lauderdale (division of Sun South). He became cagey and asked many leading questions. I said I was planning a flight to Mexico and needed 10 channels. "What frequency do you need?" he asked. The same day I went to Pan Tronics and asked for Ed Neilsson. Not there, they said. A radio tech with a foreign accent came out from the back and I asked him about getting some HF equipment, "I'm planning a trip out of the country."

"Oh, you must be the girl whose [sic] going to fly around the world." It was impossible that word could have leaked out on my flight. "Are you the girl from Ohio?" he pursued. (Pan Tronics was installing the HF in Jerrie's Ohio plane.) I played it as if it were the greatest thing since French toast.[42]

Joan maintained that at noon on January 7, 1964, she phoned the NAA and connected directly with Randleman to discuss this recent development. Randleman was an individual with whom Joan had corresponded many times before. While at first he said he had no information about any other person seeking a world flight sanction, once Joan pressed the issue, he conceded: "I can say we have received some interest from another girl a year ago. So, we'll have to make a decision about which one to sanction. We'll look at them both together." He next mentioned that he was preparing to mail Joan her sanction forms.

Joan raised a point. "But the flights are unalike. How can you measure them as alike?"

He replied, "The other flight is only for the northern hemisphere. We consider which person is better backed, and a better experienced pilot, and with a safer airplane . . . all of this we consider," he said. "Then the papers and check are sent to the FAI."

Joan persisted. "Why don't I get on the airlines and come to Washington, DC, now, fill out the papers, and give you a $150 check?" (Note: In 2019, this equates roughly to $1,200.)

---

[42] Trixie Schubert, "Joan's timeline with NAA," included as part of the original *World Flight* manuscript.

But Randleman said, "It's not necessary. Even if you're first in with the papers we'd still have to make a decision."

On January 8 at nine a.m., Joan again phoned Randleman in Miami; her bill for calls to the NAA now came to more than $300 (nearly $2,500 in 2019 dollars). She was told that Jerrie's papers had arrived in the morning mail. "It's over my head now. We can't grant your sanction now," he said. But according to *Racing to Greet the Sun*, Jerrie's paperwork was filled out in person at the NAA office in Washington, DC.[43]

In a third account obtained from Smithsonian Air and Space Museum records, Randleman did write an internal letter many months later, dated July 10, 1964, to NAA President William Ong, accounting for his interactions with the two women. While Randleman acknowledged that Joan had originally inquired about an around-the-world flight sanction in mid-December of 1963, and that she made a long-distance call to him from Miami on January 8, 1964, he did not mention anything about reaching out to Jerrie about Joan's interest to pursue a sanction in this letter, about Jerrie signing her papers in person at his office, or about Joan initially contacting him on January 7. Rather, he did explain in his letter that Joan had considered flying from Florida to Washington on the evening of January 8, assuming that another application was imminent (though Joan's account refers to this date as the January 7)[44] and that he thought it would be unfair and unethical to keep the office open for Joan after normal business hours. He also said that rather than flying from Miami to Washington, DC, Joan had decided to just fly back to Long Beach on the morning of the ninth and submit her forms via U.S. mail. Did Trixie's notes get the dates of Joan's timeline of correspondence with Randleman wrong? Or was Randleman off on his dates?

Randleman maintained that Jerrie submitted her sanction request and check for fees on January 9. From there, he said, he did call Joan back on the ninth to discuss her flight, at which point he let her know

---

[43] Phillips, *Racing to Greet the Sun*, chapter 1.

[44] Trixie Schubert, "Joan's timeline with NAA," included as part of the original *World Flight* manuscript.

that Jerrie's sanction was currently under consideration for approval.[45] Jerrie stated that she had flown to DC to sign her forms as soon as she got the call from Randleman about another woman who was looking to apply for the same sanction.[46] Joan recalled that Randleman had told her on the morning of the eighth that Jerrie's forms had already arrived by mail and that it was too late for her to submit anything.[47] While we don't have perfect information to work with, between Trixie's files, Phillips's paperwork, the Smithsonian's records, and other published accounts, one can make the observation that these dates don't all line up. Ultimately, Jerrie was granted the sanction to become the first woman to set a record for "Speed Around the World – Feminine" and "Speed Around the World, Class C-1.c (2,204 – 3,885 lbs.)"[48]

While Joan chose to press on without the sanction—she had already come so far—Jerrie's team hustled to move her departure date up by two weeks: there was a lot at stake. As both women scrambled to complete final preparations, the news media would soon pick up on their undertakings, and their circumstance would soon turn into a global news sensation. But, according to Jack Smith, it was the media that turned Joan's and Jerrie's flights into a race. In a 2017 email exchange about my interest in writing a book about Joan, Jack made the following comment to me:

> *No so-called race ever existed. The press continued to play it up, however. Joan's main objective was to complete the exact route planned by Earhart except for she would do it solo, which she did in a multi-engined airplane. Moreover, the Earhart route was around the world at the equator that was thousands of miles*

---

[45] Letter from M.J. Randleman to NAA President William Ong, July 10, 1964, National Aeronautic Association Archives, Box 147, Folder 6 (Acc. XXXX-0209). Archives Department, National Air and Space Museum, Smithsonian Institution, Washington, DC.

[46] Phillips, *Racing to Greet the Sun*, chapter 1.

[47] Randleman.

[48] Ibid.

longer than Mock's. *The band of weather is also more severe and challenging than flying around the world in the Northern Hemisphere, as was Mock's. How could two flights so different be considered a "race"?*[49]

Even though Joan and Jerrie agreed that this was never about a race, in the eyes of the rest of the world, it certainly became one. Headlines began to spring up, including, "Housewife Plans Round-the-World Flight," "Race Shapes Up: Second Woman Ready for Flight," "World Flight Set Back," "California Aviatrix Set to Go," and "They're Racing!" And then, once they took off on their respective flights, a whirlwind of play-by-play coverage ensued: "World Flight Begins Today," "Soloist Surprised in Bermuda," "Joan Merriam in Bangkok," "Globe-Girdling Women Grounded," "World Flier Faces Hard Stage," etc. The question is: did interest in both flights happen naturally, or was the spectacle created to help maximize *Columbus Dispatch* sponsorship dollars?

Winners Keepers, Losers Weepers

In the 1965 *Flying* magazine article entitled "The Loser," associate editor James Gilbert provides a thorough overview of Joan's path to flight, with a focus on her propensity for being overshadowed by bad luck. With regard to the sanctioning process, he described the whole thing as a "curious affair." He writes:

```
Then there is the curiously involved affair
of her dealings with the National Aeronautic
Association. [. . .] The NAA must have felt them-
selves in quite a spot, knowing that two girls
were thinking of attempting the same record. But
in all fairness they could have hardly said to
either one, "Psst, better get your form in quick,
because we know of someone else who may beat you
to it." All the NAA could do was sit tight and
```

---

[49] Email from Jack Smith to Tiffany Brown, November 30, 2017.

```
see whose completed application came in first. And
bad luck on the loser.[50]
```

But was that really the case? According to the dates referenced by Jerrie and Randleman, it's almost as if Randleman said exactly that to Jerrie: "Psst, better get your form in quick." According to Jerrie's account, Randleman called her directly—as a courtesy—once he realized that Joan was aware of Jerrie.[51] In Joan's account, Randleman only told her about Jerrie when pressed. From Randleman's letter to Ong, he never mentioned calling Jerrie but rather noted that he had "advised Jerrie by letter on January 6, 1964 that she should make a formal application for sanction." These statements only create more questions. Did Randleman send Jerrie a letter on the sixth, or did she receive it on the sixth? Why did Randleman send Jerrie a letter on the sixth to request that she make a formal application for sanction and not send such a letter to Joan? What prompted Randleman to send such a letter to Jerrie a full year later? Did Jerrie fly to DC on the evening of the fifth (the night before her application for sanction was signed); the evening of the seventh (when Joan says she first called Randleman and he said that he was sending her the sanction forms), or was it on the evening of the eighth (since Randleman had stated that Jerrie had submitted her forms and check on the ninth)? Regardless, if the NAA knew that two separate submissions were forthcoming, after many months of talks with each woman, then shouldn't both requests have been considered side by side?

Joan maintained that NAA had granted sanction to Jerrie even before her sanction forms had been received. In Gilbert's article, he shares what Joan told him about the affair: "I don't know how long before my flight she was aware of me, but I was first aware of her flight less than five weeks before I actually made my takeoff. You can't throw an around-the-world flight together overnight." She went on to say:

---

[50] Gilbert, p. 81-82.

[51] Phillips, *Racing to Greet the Sun*, chapter 1.

> I advised NAA of my round-the-world flight, my complete intention, route and what have you, at least five months before my flight. They told me what I'd have to do to get a sanction, and they assured me they'd work very closely with me. In December I sent all of my information for a sanction, and I did not fill out the sanction forms; but I had done everything up to that point, expecting sanction forms by return mail. They never came. On January 8th, which was three weeks after I sent my complete portfolio and life history to the NAA, I then heard about this other girl, Mrs. Mock, flying around the world. I checked with the NAA and they keep me waiting for two days and then advised me they had already granted a sanction to another girl. So you might say I tried to get a sanction but was denied.[52]

Interestingly, in the book *Three-Eight Charlie*, written by Jerrie Mock, the steps leading up to the sanctioning process are never even mentioned. Throughout her flight, speed was always the focus for Jerrie. In her memoir, she recalls that it was her husband who was pushing her to win. "He just wanted a 'first.' Not me," she wrote. Eventually, Russ and Jerrie divorced. Russ met another woman, whose newly acquired inheritance would afford them trips to Morocco and China. According to Jerrie, he fell in love with her money. "He got rid of me so he could go with her."[53]

While Joan was clearly the better, more experienced pilot, Jerrie had been working directly with the air force on her flight plan; she had solid financial backing and a better-quality airplane. In fact, the *Columbus Dispatch* was just one of Jerrie's many sponsors. And the *Columbus Dispatch* itself also happened to be part of a much larger entity known as Wolf Industries, Inc. Some of its holdings at that time included Ohio

---

[52] Gilbert, p. 81.

[53] Saunders.

National Bank, BancOhio, Ohio Company (stock brokers), and various radio and television stations. Some of Joan's sponsors, on the other hand, were a lot less corporate in nature. Key sponsors for Joan included Aztec Aircraft, Buffums department store, City of Long Beach, Dick Browning Oldsmobile, Salta Pontiac, Long Beach Chamber of Commerce, and RAJAY Corporation. Presumably, this made the decision a bit more difficult for the NAA. Or perhaps, for reasons unknown, there was never a real decision to be made? After all, Jerrie had technically inquired first. But was Joan ever treated fairly? Was there an administrative error? Had someone been rubbed the wrong way? Were there people in higher positions, with better relationships, pulling the strings? Did working with Jerrie, or promoting Jerrie's flight, offer a bigger advantage than Joan's? Or had Joan simply misunderstood a conversation or forgotten an important step in the process? Had she not been persistent enough? Or was it all just a giant, badly managed, tangled mess?

Spies in the Skies

In the 1960s, international relations were tense as the U.S. and Russia were heavily engaged in "the space race," a competition to develop superior aerospace capabilities. As each country worked hard to one up the other in the race to develop artificial satellites and unmanned space probes, as well as to become the first to achieve human spaceflight, there was a noticeably heightened level of curiosity, suspicion, and spying. For a female flying a light aircraft around the world solo in 1964, there was a real danger of being mistaken for a spy while flying through or near a country's restricted airspace. As such, both Joan and Jerrie took extra precautions in their planning to avoid potential disasters.

According to the book *Operation Overflight: A Memoir of the U2 Incident*, beginning in 1956 through 1960, the U.S. was spying on not only most of the countries in the Middle East, but also its allies.[54] The book's author, American pilot Francis Gary Powers, would know: in May 1960, his CIA U-2 spy plane was famously shot down by a Russian

---

[54] Francis Gary Powers, *Operation Overflight: A Memoir of the U2 Incident*, (Lincoln, Nebraska: Potomac Books, December 1, 2003), p. 262.

surface-to-air missile. Subsequently, he was tried very publicly in Moscow and sentenced to prison for espionage. In the book, Powers gives a thorough overview of the events leading up to 1960, and he talks in detail about his involvement with executing secret overflight missions on behalf of the U.S. government.

As it turns out, Trixie was in Moscow and on the courthouse steps during the Francis Powers trial in 1960. Not only does her personal account of traveling behind the Iron Curtain offer some unique insight into the reality of the Cold War tensions experienced during this time, but it speaks to the nuances of spying. Having traveled into the country with twenty-six high-ranking, out-of-uniform military officials on a German sightseeing bus, Trixie ultimately learned that posing as a housewife on a sightseeing trip was the best way to gain access to information inside the country as a journalist. Though she had earned her own pilot's license in advance of this trip, she would never need it given the sensitive relationship between the two countries and their tightly monitored airspace. During a lecture she gave to the American Association of University Women in 1964 about her time behind the Iron Curtain, she reflected:

> My first opportunity overseas with a journalist passport was to ask for a visa to go into Moscow. It was denied both by the Russians and the Americans. By the Russians, on the grounds that they had Francis Powers, and he would be tried. They had more journalists than they wanted in the country. The Americans turned down the visa on the grounds that I had a brother in counterintelligence in Turkey. And with three small children, I would be an apt hostage should they decide to keep me. So I went to Heidelberg, where the new state headquarters of the American Military is located. Hand carried my papers and didn't fill them out quite as adequately as I had with the Air Force. I said I wanted to go to Moscow as a sightseer, and my occupation was, "housewife." My visa was granted.[55]

---

[55] Trixie-Ann Shubert, "Before and Behind the Iron Curtain," lecture given to the American Association of University Women in Whittier, CA, (specific date not known) 1964.

Upon boarding the bus in Germany that would travel through Czechoslovakia and Poland to Moscow, Trixie was handed a list of one hundred fifty things she could not photograph. She was told not to speak the language and not to ask too many questions, rendering a "sightseeing" trip impossible. Throughout her lecture, she shared details about sights she saw as an American visitor to the region, the things that stood out to her about traveling through a Communist country, and the nerve-racking experiences their group faced along the way. While she was not technically a spy, she too was collecting insights while working closely with the U.S. government. In other words, she was a convenient asset.

At the same time that Trixie was reporting from behind the Iron Curtain in Russia, San Francisco-based CBS radio broadcaster Fred Goerner was investigating the Amelia Earhart disappearance for himself on the island of Saipan, where Amelia Earhart and Fred Noonan were rumored to have been held captive in 1937—coincidentally, for spying. Goerner's initial visit to the island in 1960 had been prompted by meeting with a woman living in San Francisco, originally from Saipan, who believed that Amelia Earhart had been on the island after she disappeared. In his 1966 best-selling book *The Search for Amelia Earhart*, Goerner shares the research he uncovered in support of the idea that Amelia and Fred were captured by the Japanese following their disappearance and were taken as prisoners to Saipan. Upon his book's publication, he would also expose a shocking secret: the CIA had been operating a covert spy-training facility in Saipan under the cover of the U.S. Navy from approximately 1953 to 1962.[56]

With respect to Joan's desire to complete the Amelia Earhart route, upon reading this piece about the CIA's spy-training facility, I immediately wondered: when Joan submitted her full route itinerary to the NAA in 1963, could the CIA have become involved somehow? I thought back to the January 1963 letter that Russ had received from Randleman explaining what would be necessary for Jerrie to set an aviation record for "Speed Around the World." In it, Randleman mentions that the FAA

---

[56] Fred Goerner, *The Search for Amelia Earhart*, (Doubleday and Company, Inc.: Garden City, NY), 1966, p. 115-126.

wanted to be advised well in advance of all extended foreign flights in light aircraft in order to be of maximum assistance in planning, as well as to evaluate the safety of such flights. Since Joan's intention was to complete the Amelia Earhart route through New Guinea, Saipan, and elsewhere in the Pacific, could her flight have been too much of a potential public relations disaster for the federal government?

Mike Campbell, who wrote the 2012 book *Amelia Earhart: The Truth at Last*, served as an award-winning print and broadcast journalist while on active duty with the U.S. Navy, and as a civilian public affairs officer with the air force. On his blog, *Amelia Earhart: The Truth at Last*, he explains the significance of the Naval Technical Training Unit (NTTU), the spy-training facility that was at the time called the Saipan Training Station:

> In a July 1961 memorandum from Brig. General Edward G. Lansdale, Pentagon expert on guerrilla warfare, to Gen. Maxwell D. Taylor, President John F. Kennedy's military advisor on Resources for Unconventional Warfare, SE Asia, Lansdale wrote: "In 1948, the Central Intelligence Agency (CIA) closed off half of Saipan to islanders and outsiders, using the northern part of the island for covert military maneuvers. The end of WWII left a power vacuum that was being filled by Communism; the Cold War objectives of the CIA's covert facility on Saipan were to thwart communist expansion. The island's remoteness and control by the military made it an ideal base for this training and the NTTU was established. The primary mission of the Saipan Training Station [was] to provide physical facilities and competent instructor personnel to fulfill a variety of training requirements including intelligence tradecraft, communications, counter-intelligence and psychological warfare techniques. Training [was] performed in support of CIA activities conducted throughout the Far East." The NTTU was established in 1953 and closed down in 1962. Fred Goerner wrote at length about the

NTTU and its role in discouraging media from visiting Saipan in search of Amelia Earhart.[57]

While Trixie may or may not have not known about NTTU in 1960, it would have been well established during the period when Trixie was traveling through Europe. Had she known about its existence, Joan's dedication to traveling in that region may have raised a red flag for her. At any rate, it would not be a stretch to say that the FAA, the NAA, or, for that matter, the CIA, would have been interested in any female who wanted to be the first to achieve an around-the-world solo flight in 1964 through this particular region—especially one who wanted to publicly retrace the steps of Amelia Earhart and investigate her disappearance, which was sure to reignite interest in her case.

An Added Lift?

According to *Racing to Greet the Sun*, Jerrie worked with the air force in charting her flight stops. And, immediately following the in-person submission of her sanction forms in Washington, DC, she had an appointment at the Pentagon.

> A year had passed since she had first visited the office of Colonel Lassiter at Lockbourne Air Force Base. During that time, he had become General Lassiter, stationed in Washington. General Lassiter picked Jerrie up at the NAA office and gave her a ride across town in his limousine. She was escorted into the heart of the Pentagon and allowed a glimpse of a computer the size of a bedroom while the best military strategists in the world helped determine a safe route for her. Jerrie sorted through all the regulations and contacted all the foreign embassies to make certain that, in this Cold War era, no government would think she was a spy.[58]

---

[57] Mike Campbell, "White's Earhart story another in long Saipan list," *Amelia Earhart: The Truth at Last* blog, published February 23, 2017, https://earhart-truth.wordpress.com/2017/02/23/whites-earhart-story-another-in-long-saipan-list.

[58] Phillips, *Racing to Greet the Sun*, January 19, 2015, chapter 1.

To be clear, charting a world flight and obtaining international clearances were tricky: to plan accurately would have required some level of federal government assistance. Amelia Earhart herself spoke about the complexities of the process in her book *Last Flight* and how she worked with the state department to accumulate proper permissions, paperwork, and clearances ahead of traveling through various regions, noting that some countries even required "a testimonial of character and a negative police record" to pass through.[59]

But what's interesting about Jerrie's meeting with General Lassiter is not so much the fact that he helped her chart flight stops, but, rather, that in the very same year, General Olbert F. "Dick" Lassiter would be named president and chairman of the board for Executive Jet Aviation (EJA). Today known as Netjets, Inc., EJA, which was headquartered in Jerrie's hometown of Columbus, Ohio, became the first private business jet charter and aircraft management company in the world. There were likely connections between EJA and the Mocks—Jerrie being a pilot and Russ working in advertising. Is it possible that Lassiter wanted to help Jerrie for personal reasons, or was this simply the result of a friendly connection stemming from a level of mutual respect? In his 2001 article in *Wired* magazine about the history of Netjets, writer Warren Berger says:

```
In the early 1960s, retired Air Force general
Dick Lassiter told actor Jimmy Stewart (whom he'd
gotten to know during a stint as a technical
adviser on war movies) about his plan to launch a
private jet charter service for powerful business
executives and celebrities. Stewart agreed to kick
in money, as did TV star Arthur Godfrey. Another
well-known military man, General Paul Tibbetts
(captain of the Enola Gay), got involved, too.
This crew launched Executive Jet in centrally
located Columbus in 1964, and promptly bought the
```

---

[59] Earhart, p. 44.

```
first Learjets that rolled off production lines.
Executive Jet became the pioneer of private jet
chartering, known for its fast planes and attrac-
tive stewardesses (Lassiter was said to have had
a taste for both).⁶⁰
```

We also know that, in addition to working with Lassiter and the Pentagon, Jerrie was visited in person by the CIA following her around-the-world flight and that photos of her trip went missing and were believed to have been rerouted to the CIA.[61] Clearly, the CIA was interested in what Jerrie was up to, even though she had taken all of the necessary precautions. Could the CIA have been using Jerrie's flight to monitor international conditions as she encountered them around the world, even if unbeknownst to her? Could the CIA have been paying close attention to Joan's flight as well, even though she had not been sanctioned? A bored American housewife with small children who was flying around the world, after all, would certainly have been a less worrisome public relations issue than the wife of a naval lieutenant stationed in the Pacific whose life goal was to bring justice to Amelia Earhart's unfinished attempt to circumnavigate the globe.

Preparing for Takeoff

Upon discovering that there was now an unplanned and unexpected competition taking place, Joan undoubtedly realized the importance of rallying her troops to secure support prior to embarking on her journey. Before setting off for her around-the-world flight, she wrote a letter to the International Headquarters for The Ninety-Nines, Inc.—nearly a year before her fatal plane crash. Here's an excerpt from her February 19, 1964, letter to the Ninety-Nines:

---

[60] Warren Berger, "Hey, You're Worth It (Now)," *Wired Magazine*, June 1, 2001, https://www.wired.com/2001/06/netjets/.

[61] Phillips, *Racing to Greet the Sun*, epilogue.

Dear 99s:

This "long" letter is to announce my plans to you of my forthcoming "Around the World Solo Flight," which will begin in about 30-45 days. As per the enclosed itinerary, you will note that this is the same route as the 1937 A.E. route. I have had this secret ambition, not only to attempt to be the first to fly around the world, but when I did I would like to fly the Earhart route after reading the history of the same for so many years.

I would like the support of the 99's during this flight and would appreciate any assistance that you might lend after the flight, publicity wise. This is not a stunt, but rather a carefully planned flight which is very close and important to me. Planning has been constant for 14 months, and the desire to accomplish the same has been with me since 1955. Due to the sponsorship involved in the flight, for which I am very grateful and without same this flight would be impossible, it is necessary for me to publicize the flight before, during, and afterwards.

A turbocharged Piper Apache will be used in the flight, which is well equipped with radio and new engines. My experience is outlined in the enclosed resume also. Many improved radio aids enroute exist today, and with the higher performance of today's aircraft and reliability of modern radio equipment, a flight today certainly has a thousand times the advantage as did our first president, Amelia Earhart, in her last flight.

To my knowledge another girl will be attempting to beat me at this in a single engine Cessna, along a shorter route to the north of the Amelia Earhart route. She is not a 99, lives in Ohio and probably will be departing about the same time as I will. She made an early press release January 31st, and mine will be out this week, so I did want to make sure that you were aware of the two girls attempting this.[62]

Realizing the need to garner publicity for this one-of-a-kind adventure, Joan also enlisted the help of a public relations professional—a veteran Southern California-based advertising executive by the name of John Sarver with Sarver & Witzerman. His purpose was to help promote her efforts throughout her flight.

And with that, just a few short months later, on March 16, 1964, twenty-seven-year-old Joan was sitting on the runway in Long Beach preparing to take off for Oakland, California, on the first leg of her around-the-world journey. The official date for takeoff from Oakland would be March 17, 1964—the very same date and runway on which Amelia Earhart had taken off exactly twenty-seven years earlier.

---

[62] Letter from Joan Merriam Smith to The Ninety-Nines, Inc., February 19, 1964.

PART TWO

A WORD FROM THE AUTHOR

Now that you're familiar with why I became interested in this story and how we got to this point more than fifty years later, it's time to hear more about what Joan accomplished, and in her very own words. What you are about to read contains the transcribed version of the original manuscript that I came across, in the condition in which it was presented to me, after long ago being stored away in a box with my grandmother's things. Because the version of the manuscript I had in my possession was more like a final draft, littered with handwritten notations and comments throughout, I did my best to capture every word and statement accurately. Since I had a fair amount of difficulty translating various pages that were nearly impossible to read, whether that was due to faded text or illegible notes, I simply omitted those areas and focused on bringing you the information that I did have.

Trixie compiled material for her book—*World Flight: Joan Merriam Smith*—based on information provided directly by Joan. It is my understanding that she relied heavily on the tape recordings taken by Joan during her world flight, and that she used that information as a basis for how this story was presented. As such, the manuscript is written in Joan's first-person voice. As indicated earlier, Trixie's personal journals underscore the fact that she had regular meetings with Joan to discuss content and that those meetings were likely used to help Trixie embellish and build upon observations and reflections made in Joan's own recordings. It appears she followed a similar approach while writing *A Bell in the Heart: The Autobiography of Patty Gardenseed,*

America's Ambassador of Good Will, in 1963. That book was written from the first-person perspective of navy veteran Aloysius Eugene Francis Patrick Mozier, who went by the shortened name Patty Gardenseed.

Based on Trixie's system of numbering at the top of each page, the version of the manuscript I translated was clearly missing chapters, most notably the opening chapters (1-4) and the final chapter (16). But thanks to that numbering system, I was also able to easily identify missing pages from within the chapters, although those pages were very far and few between.

To account for the missing information, and to make better sense of the information that I did have, the first thing I did upon transcribing the content was to divide the full text into subsections based on dates referenced. In this way, I was able to make sure that the story I had followed a linear order. From there, I subdivided the story again into three main parts, according to the section of the world that Joan was traveling through at the time. In other words, while my transcribed version closely reflects the text as it was intended to be written from Trixie's original manuscript, her structure would have been much different, with sixteen original chapters. Her opening and closing chapters would have also featured thoughts and reflections that we'll likely never be able to know.

Due to the various missing chapters and pages, I had to research deeply to ensure that I was bringing you the most complete and uninterrupted story. My goal for wanting to piece this full story together, after all, was what ultimately launched the research that would became the basis for this book. As for the pages that were missing from the original copy, because I was able to locate many original writings from other publications completed by Joan pertaining to the experience of her flight, in just a few places, I have bridged Trixie's narrative together with information supplemented from other articles, for the purpose of continuity. For these portions, you will find footnotes and quotations directly sourcing the locations of that original content. I did my best to present any missing gaps to you in a manner what was still faithful to Joan's own voice yet also accurate.

Wherever possible, I maintained Joan's original words. This means that in many places, I opted not to edit her language, despite various grammatical inconsistencies, outdated phrases, misspellings, and shifts in tenses. She had a unique voice, and it was my intention to capture it, so please forgive any editorial errors, as I felt it was important to retain as much as possible of this historical document that was entrusted to me.

= CHAPTER 3 =

THE GREAT SEND-OFF: FROM OAKLAND TO MIAMI

Joan took off from Long Beach bound for Oakland on March 16, 1964. Her send-off was a joyful event: city officials, members of the press, the Long Beach Municipal Band, and many friends joined together to bid her goodbye on a flight that would become the grand adventure of a lifetime. Recalling the excitement in the air on the morning of Joan's departure, an individual —with whom I originally connected via my blog post—shared his reflections: "I remember what the air in Long Beach felt like, how the Apache looked painted with many logos and script, how the band sounded, and the excitement on the ramp the day that Joan departed for Oakland. 'That's our friend, Joan!' I thought, with the excitement of a child."

Flying through some of the roughest weather she had ever encountered at that point in her flying experience, Joan's takeoff from Long Beach for Oakland was so windy that gaining altitude was difficult, and it would be an hour before she could turn on her tape recorder due to being tossed all about the sky. This would ultimately set the stage for the treacherous trip that lay ahead. Because of the fact that Joan was following the equatorial route over some of the most remote regions of the world, her flight would not only be 4,650 miles longer overall than

Jerrie Mock's, but she would face more inclement weather, inferior access to plane service, and less predictable radio communications.

When Amelia Earhart first embarked on her world flight on March 17, 1937, she was headed from west to east, traveling from Oakland to Hawaii. It wouldn't be until her plane broke apart on the runway upon takeoff in Hawaii on March 20, 1937—the second leg of her first around-the-world attempt—that she would need to cancel the flight and start over again. Three months later, on June 1, she would again take off from Oakland, this time switching directions, traveling from west to east. This eastbound route was the same route that Joan would follow.

Later that morning, Joan arrived at the Oakland airport from Long Beach, where she was cheerfully greeted by the media. According to her public relations representative, John Sarver, Oakland went all out to celebrate her arrival. Then, on the morning of Tuesday, March 17, Joan took off from Oakland at exactly one o'clock p.m. to officially kick off her around-the-world trip. Again, the full contingent of members from the press, various officials, and friends in light planes followed her out for about thirty minutes. She arrived safely in Tucson, Arizona, after a good flight, but the weather was building up. Due to more incoming weather, Joan had to take off at two a.m. on March 18. The weather became so bad that she had to dodge it, under the direction of radar ground forces. Due to a diminishing gas supply, she was forced to land in Lubbock, Texas. She continued to fight the weather en route to New Orleans.

On March 19, Joan left New Orleans, attempting to keep ahead of the weather front that was bearing down, and headed for Miami, arriving there about noon. The City of Miami hosted a grand celebration to welcome Joan. Miami, after all, was the city where Joan had lived for many years and also where she had learned to fly. She stayed for two days in Miami as she had planned to do, getting last-minute plane checks and picking up her emergency overseas equipment.

The following begins Trixie's manuscript, in which Joan shares her first-person account of the flight.

March 16, 1964: Shakedown Cruise to Oakland

Five One Poppa, red, white, and clean, looked trim and sky worthy on the flight line at Long Beach airport. She was as ready as she'd ever be for the shakedown cruise to Oakland, start of the equator flight. Across the upper fuselage were blazoned the words "Around the World Solo Flight, 1937 Amelia Earhart Route." On her nose spelled out "City of Long Beach." She was so named in deference to my adopted city, which was a partial sponsor and ready to give me a magnificent sendoff. At first I toyed with calling the plane "Amelia II."

Amelia Earhart, apart from receiving her plane from Purdue Research Foundation, received substantial sums of money for her venture from the Floyd Odlums (Jacqueline Cochran)[63], Admiral Richard Byrd, and Bernard Baruch. I felt it a courtesy to Long Beach, one of my major sponsors, to name the plane "City of Long Beach."

Paul Mantz, who was Amelia Earhart's technical advisor, offered me similar service. Paul, only person to win the Bendix races three times consecutively, is a colorful stunt pilot who currently devotes his energies to restoring and preserving one of each type of aircraft in his West Coast air museum.

All was in final readiness on March 16. The Long Beach municipal band was on the improvised platform to send me winging on my way with music. Mayor Edith Wade, California Ninety-Nines, TV, press, and radio, mechanic crews from Aztec Aviation who had worked on my plane, and a crowd of well-wishers waved me off as 51-Poppa skimmed down the runway and lifted off. The three-hour flight to Oakland was rough, turbulent, living up to its name literally—shakedown cruise. From Gaviota I cut over to Santa Maria, changing altitude from 4,500 to 8,500 but not escaping the roughing up from the strong northeast winds. The headwind, where usually there are prevailing westerlies, was a portent of the headwinds that would plague me 90 percent of the route around the world. It had taken an hour to load the plane. Not that there was so much to load, but that it had to be done systematically, utilizing every square inch to advantage.

---

[63] In 1932, Jacqueline Cochran, an aviation pioneer, married millionaire financier Floyd Odlum.

Where was the advantage? There was too much stuff around my feet. I was cramped and uncomfortable from the start. The line from the old song "five-foot-two, eyes of blue" fits me precisely. Though I'm small, the space in the cockpit is not sufficient.

I would make changes in Oakland, eliminating something somehow, something not vital. When Amelia Earhart lightened her load by leaving behind 250 feet of antenna before the fatal last leg of her flight, she cut a critical lifeline to survival over the sea where homing devices were nil, according to some of her biographers.

This was as good a time as any to try out the special Noreloo recorder on which I would record sights, sounds, and impressions around the world. Amelia Earhart kept a written log of her flight and sent sections of it back to the States after completing each leg along the route. I would send back tapes after each leg along the route. The recording device was small, compact. I turned it on, talked into it, noting that everything in the plane appeared to be working normally. Two p.m. and abeam Vandenberg Air Force base at 6,000 feet. The ocean has quite a chop to it. Now abeam Castle Air Force Base. Every time I pass Castle I look down at the jets there and think that someday I'm going to visit the base.

I'm on my way at last after all the planning and preparation. I'm on my way. It is good, for a change, to be alone. Up here is another world and I belong to it.

I've been listening to weather broadcasts, always on schedule 15 minutes before the hour and 15 minutes after the hour with special reports for pilots wherever deemed necessary. Like today when they're broadcasting severe turbulence warnings for light aircraft. These reports are part of the service of omni stations[64] around the world. With Oakland in sight and permission from the control tower, I put down on runway 27 Right, almost directly into the sun. There appears to be a crowd of people on the field. I will have to learn, even when tired and irritable, to treat the

---

[64] Very High Frequency (VHF) Omni-Directional Range (VOR) is a type of short-range radio navigation system for aircraft, enabling aircraft with a receiving unit to determine its position and stay on course by receiving radio signals transmitted by a network of fixed ground radio beacons.

well-wishers and the press people judiciously, with courtesy and tact. And I shall hope that friends and public will understand why, in the interest of sleep and alertness, I'll be declining receptions, parties, late hours until the world flight is accomplished. Tonight, for instance, the eve of my departure, I want only sleep, rest, and thoughts of the flight.

March 17, 1964: The Official Send-off

March 17 dawn broke fuzzily through the California coastal fog to usher in St. Patrick's Day, 1964, the 27th anniversary of the start of the Earhart-Noonan flight. Amelia Earhart's first attempt at the world flight had been March 17, and westbound because the weather was better that way. When the plane cracked up on takeoff in Hawaii, the flight was deferred until May when her second and final attempt took off eastbound. By an interesting coincidence the first world flight of U.S. Army fliers in 1924 also took off on St. Patrick's Day, March 17 from California.

Early this morning I bought a smaller suitcase, repacked everything in the plane, and had all readiness one hour before the press arrived so that I'd have no last minute worries and could devote my attention to answering questions and being gracious. Mayor Houlihan presented me with the keys to the city of Oakland, the first of many such honorary keys.

Right from the start there were parallels to the Earhart flight. My plane was hangared in number 22, the same hangar which had housed her plane for the night. I was cleared to take off from the same runway she had. And airport manager Fred McElwain, who had handled Amelia Earhart's flight 27 years earlier, handled arrangements for mine. Elmer Dimity, now aged and on crutches, came up to wish me luck. He was a friend of Amelia Earhart and helped me raise funds.

I climbed into the cockpit. The earlier fog had dissipated into a haze but hung over Oakland. My thoughts were whirling through outwardly. I know I appear very calm. How many thousands of takeoffs made. And yet none have been, or ever again will be, quite like this one. How to describe it.

In a strange way, going down the runway with 51-Poppa had some of the finality, the adventure, the for-better-or-worse-but-forever quality of going down the chapel aisle with Jack. This was it. There was no retreating. The future was committed.

Precisely at 1 p.m. 55-Poppa and I were airborne from runway 27 Left. The control tower permitted a left turn out and I climbed on course toward Bakersfield and Tucson.

The world flight had begun. I noted that I had an escort of small planes: Tom Hudson, flying off my left wing in an Aztec, a Cessna 310 and another Aztec carrying airport and city officials, and Fred Goerner (whose interest in this flight was more than casual) in a Sky Knight trying to slow down sufficiently to keep up with me and finally turning back. Fred's concern about Amelia Earhart's disappearance had taken him several times to Saipan, under the auspices of CBS news, to search for clues to her disappearance.

I turned on the recorder. The Sierras are on my left. Snow crests the mountains on almost all peaks. Off to my right is Monterey where Jack and I used to live. I have a koala bear as a good luck charm. Tom Hudson and other friends had him made for me by Peter Carlyle who is also a pilot. They presented the little koala to me on behalf of the Monarch Aviation from the Monterey Chamber of Commerce. They tell me Australian pilots carry such a good luck bear. He's made of kangaroo fur.

Two hours out. Indicated air speed 130 knots. That's about 160 miles per hour ground speed. I'm allowing five hours as my estimated time en route to Tucson. There were so many people at Oakland, some of them longtime friends. I had no time to talk with them before takeoff. Already I began to feel some of the sting of being in the spotlight. One is bound to offend some people unwittingly.

Blythe, California, lies just ahead and the high winds have out visibility to five miles. Even though I rearranged the cockpit I still have trouble shifting around in these small quarters. My right foot is asleep. The radio reports some weather developing northwest and northeast of Tucson. I'm still operating on the ferry tank, the cabin tanks inside the cockpit where there used to be four additional seats in this five-seat airplane.

Just after dark I arrive in Tucson and touch down alongside of the handsome new terminal building, quite ready for the coffee the waiting Ninety-Nines brought me. But I was concerned with the weather reports and drank the coffee on my way to the weather bureau. A low pressure area was in the making, probably icing on the wings of the plane, and turbulence again. I would have to leave again as soon as possible, without the night's sleep I was looking forward to.

In any flight, whether it's from Oshkosh to Manitowoc or from Oakland to Oakland via the equator, the contact crew and women on the ground are the flier's vital link to safety to their final destination. The personnel who staff the control towers of airports around the clock provide the landing and takeoff instructions, wind direction, altimeter settings, area traffic to the pilot; the FAA radio man aids remote stations and makes certain that the electronic guidance equipment is functioning to keep instrument pilots on course and answer their calls for assistance for weather and special warnings of turbulence. The meteorologists who know the importance of weather wind velocities and directions at all altitudes up to date constantly plot high- and low-pressure systems, pressure gradients and never are too tired or too busy to scrutinize the weather charts and data with pilots requesting help.

I'm grateful for the help and cooperation of control tower personnel, omni stations, meteorologists who have been alerted that this is not just pleasure flying, but globe girdling, and want it to succeed. Already I had reason to be grateful for added considerations from aviation people. Albuquerque had a B-52 Air Force plane up getting reports on winds aloft for the Air Force. The information was passed along to me. Pilot reports, too, were being expedited to me via aviation channels. Thunderstorms were building southwest and northeast of Tucson which meant I'd have to deviate north to get around the rugged weather at night.

The people at Tucson had plans to celebrate my arrival. I could expect the Ninety-Nines to understand why I had to push on. But how to explain to city officials that the welcome festivities would have to be whittled down to dinner and a quick departure.

There often is the danger in this Tucson area of a low pressure system forming and moving east/northeast where it picks up Gulf moisture and spreads into the Gulf States (Arizona and New Mexico), pushing north. So many times this creates a wide area of weather over half the United States. The more moisture from the Gulf, the worse the weather.

I leave the weather bureau long enough to sit down to a chicken dinner and a cake molded into the shape of a plane. It can't be by chance that the pianist in an adjoining cocktail lounge was playing "Around the World" and "Come Josephine, in My Flying Machine."

Back in the weather bureau I noted that El Paso and Deming were picking up cloud cover and some rain, meaning the low pressure was forming and deepening. Obviously I had to get out, and fast. El Paso had a 4,000-foot ceiling. The airport elevation there is 4,000 feet, so the clouds were 8,000 feet above sea level. You scarcely can clear terrain in some of that area at 8,000 feet. Because of the cold temperature there would be ice in the clouds, the kind of weather I've experienced often when flying in mountain areas. There isn't too much forecasting available on this type of low pressure because it forms off the Pacific Ocean and moves in so that no past forecasting is available on which to base prediction for the immediate future. It amounts to the start of a new weather condition, and trying to forecast is mostly guessing.

March 18, 1964: Tucson to New Orleans

It was midnight. I had hoped to be deep in sleep at this time instead of worrying out the weather in the weather bureau. I couldn't get out of Tucson to the east or northeast, so I would have to go north through Albuquerque, even though the minimum en route altitude is pretty high, 11,000 feet.

I had a final cup of coffee, filed a flight plan, and was in the air again at 2:30 in the morning on the 18th. I flew north to Phoenix, then to Winslow and east to Zuni and Albuquerque. Tucson departure control and radar control talked with me quite a bit as there's no military or airline traffic in the air at this hour. I enjoyed the chitchat as it helped keep me awake. One hour out of Albuquerque and the weather was already moving in. I hoped I wouldn't get into Albuquerque before daylight as I wanted to see the weather. No horizon. It was pitch black. There are no town or city lights below. I was flying strictly on instruments. The glare of the red instrument lights, plus the wing tip lights, and the swish of the rotating beacon red light against the wing can cause vertigo. They make me sleepy. I'm very tired anyway. I switch on the oxygen regularly to keep fairly alert.

Albuquerque forecast is not too good so I push further north to stay out of the weather. I land at Lubbock, Texas, to refuel as a precautionary measure. One hundred miles further west, just west of Fort Worth, I run out of luck. There is weather to the ground. No icing conditions, however. The freezing level is usually lower in the mountains because one must fly at a higher altitude.

I obtain instrument clearance through to Shreveport and New Orleans and remain on instruments for four hours. Not too much turbulence. I arrive at New Orleans' Moisant Field a little after 4 p.m., having broken out of the weather at Baton Rouge and flying on visual flight rules the remainder of the way. It was then that I began flying a little lower and picked up a little speed. Suddenly I heard a loud crashing noise. Did the baggage door open? Was it the landing gear doors? Antenna loose? Or could it be the rubber stripping blowing off the metal fairing around the wing? The stripping underneath the wing is a streamlining device geared to cut down somewhat on the drag. I called the approach control but made it clear that I was not declaring an emergency. I met another aircraft which flew alongside and saw some long black object flapping against the bottom of the plane.

As I landed at New Orleans four or five trucks met me. My situation had been labeled an emergency even though I had not indicated an emergency. Pilots must file a special report with the FAA if declaring an emergency landing. The problem was that I finally had deduced the long rubber stripping of fairing was flapping against the plane. The decision was made to simply cut it off rather than delay with repairing it, as it was not a vital function of the plane.

On for a blessed night of sleep. Lillian Huber, a commercial pilot and instructor who learned to fly at the same time I did, meets me and takes me home with her. Lillian's husband is a Delta Airline pilot. I sleep almost at the moment my head touches the pillow, but not without first checking weather. There is a forecast for the same conditions east of here as I found in Tucson. So I get up at 3 a.m., the 19th, am at the airport by 4:15 and in the air at 4:55.

March 19, 1964: New Orleans to Miami

Thunderstorms. The radar center is working me around these storms and advising headings. I take a 100 degree heading over the center of

Lake Pontchartrain and call Mobile [airport] to get a clearance across the Brookley Air Force Base. I need a clearance also through the Tyndall Air Force Base area around Panama City; then I can fly more directly and save time. This area brings back fond memories of when I worked for the West Florida Natural Gas Company at Panama City, and where I met Jack.

I turn on the turbochargers to increase speed. As I go over Panama City I'll give a call to Bill Sowell at the field to say hello. I used to keep the company Apache in his hangar.

Radar has handled me from unit to unit on this leg. One unit handles you maybe 100 or 150 miles and then passes you along to the next radar screen. This method is called a radar handoff. The Alabama River is up the left now. Still another air force base area, Eglin, requires clearance through a restricted area about 100 miles long. There is a parachute jump training as well as other military maneuvers through here. The weather is pretty good today. One has to be in the air early to beat the showers that concentrate along the Florida Gulf Coast, caused by the high humidity in the air.

I glance out at the signatures decorating 51-Poppa. I can see the names of American, Delta, and Eastern airline pilots and personnel. Just before takeoff, at 4:30 this morning, Lillian and I found two teenagers signing my plane. They seemed to be quite thrilled when we arrived unexpectedly on the scene. They said they couldn't believe such a small plane was going around the world. It seems only yesterday I was a teenager myself, awestruck with aviation.

I make a 20 degree wind correction, necessary to make good my track across the ground. I've been smelling gas fumes to the left of me. They probably were spilled last night over the top of the tank. If gas leaked down underneath and the odor became stronger it could mean danger. I'd have to shut my entire electrical system to avoid a fire. I forgot my medicine kit this morning at New Orleans; I will have to try to radio a station there so it can be air freighted to Miami. It has vitamin pills, things I might not be able to get at some of the stops around the world. I have Miami International in sight. Flight time 5 hours and 30 minutes. Arrival time 11:45 a.m. I shut off the recorder, the radio, the engines, the master switch, in that order.

Mother and Aunt Beryl and Uncle Frank were there to meet me at the end of this final transcontinental leg of my route, beaming. Mother lives only three miles from the airport. It was difficult to even my mother that I'll have very little time to spend with her this trip as there is much to do before my departure on the 21st for San Juan, Puerto Rico. Time only for the briefest of visits. There was too much else on my mind. Like getting my survival gear from Pan Avion.

I was taking a two-man life raft, life vest, and survival equipment for desert, water, and jungle. Within easy access in the cockpit I stacked 28 Manila envelopes with the 105 charts I needed to navigate, a supply of Life Savers candy, a sextant[65] and tables for sunline navigation, a 22-calibre gun, a couple thermos bottles, my 33-mm camera, the tiny tape recorder, passport and visas, St. Christopher medal on the plane upholstery, two pairs of sunglasses, $3,000 in traveler's checks, and the small suitcase with three changes of wash-and-wear clothes, the omnipresent extra gas tanks which will dominate the cockpit for the duration of the world flight.

I'm wearing only dresses on this flight, not slacks. I guess I'm pretty well indoctrinated with the Powder Puff Derby philosophy that women need not be unfeminine to fly. So I have a couple drip-dry dresses, a couple skirts and blouses, all wash-and-wear type. It will suffice.

Although the plane is insured I have no personal insurance on this flight. What money I could get together is needed for the plane. And anyway, Jack, who would be my beneficiary, says if anything happened to me, he couldn't bring himself to accept to spend the money anyway. So that settles that.

---

[65] A sextant is a doubly reflecting navigation instrument that measures the angular distance between two visible objects. The primary use of a sextant is to measure the angle between an astronomical object and the horizon for the purposes of celestial navigation.

= CHAPTER 4 =

BEARING DOWN ON BRAZIL, EXPLORING AFRICA

AFTER PACKING SURVIVAL GEAR INTO HER plane, Joan took off from Miami early on March 21, 1964, later arriving in San Juan, Puerto Rico. On March 22, she departed from San Juan for her second leg across water and arrived in Paramaribo, Suriname—a former Dutch colony on the northeastern coast of South America—where some magneto trouble was immediately corrected. She noted that the town fit Amelia Earhart's description of it precisely: "a town squatting at the edge of the Surinam River, lots of red-tile rooftops and some thatched huts."[66]

After a routine overnight stop, Joan was ready again for takeoff on the morning of March 22, having filled her tanks the night before. She slept at the airport's VIP quarters, which in actuality was just an upstairs room. As she was filing her flight plan, the locals called her, very excitedly, for the tanks were leaking profusely at the welded seams, and gasoline was pouring out of the tanks at a rate of about two gallons a minute.[67]

Up until this time, the entire flight had had its problems, but everything had been right on schedule. Per Joan's press agent, John Sarver:

---

[66] Joan Merriam Smith, "I Flew Around the World Alone," p. 78-79.
[67] John Sarver, "Sarver & Mitzerman Press Release," April 9, 1964, courtesy of 99s Museum of Women Pilots.

> *Joan immediately informed me by phone of this delay, and a more disgusted girl in this world I have never heard. But, she was absolutely determined to repair the tanks and continue. The tank installation is quite complicated and it is a six-hour job just to take them out or put them back. Joan had spent thousands of dollars to get the very best "ferry tank job" possible, and this was the least expected trouble.*[68]

According to Joan, if a tank had cracked in flight, it could have been disastrous, causing either explosion and fire or asphyxiation from fumes. Reflecting on her time spent on the ground in Suriname, she would later recall:

> *I pushed everybody as hard as I could, but Surinam [sic] was pretty primitive, and I was really afraid I was going to have to abandon the trip right there. The tanks had to be shipped 30 miles from the airport to Paramaribo for repairs. Everything moved at a maddeningly slow pace. Days passed, and the rains came down. I stayed right at the airport most of the time, standing by the plane, drinking gallons of black coffee and wondering what this would do to my carefully planned flight schedule.*[69]

Finally, after the repeated welding and rewelding of tanks and six discouraging, heartbreaking, and frustrating days for Joan, the tanks were fixed as well as they could be, wrapped in plywood, and held together with metal banding for needed reinforcement. Gas had filtered through all of her radio equipment.

On March 30, 1964, Joan took off for Natal, Brazil, which is located at the northeastern tip of the country. Dodging thunderclouds, she had to fly underneath them at about five hundred feet, fearing that the sky might collapse on her at any given time. She recalled, "One thought that never left my mind: those alligators. At 300 feet I could see the Amazon

---

[68] Ibid.

[69] Joan Merriam Smith, "I Flew Around the World Alone," p. 79.

jungle and swamps in detail, and I knew if I were forced down here, my chances of getting out alive were awfully slim."[70]

Fierce weather conditions at the equator would next prompt another forced landing, this time in the heart of the Amazon jungle. Upon seeing the primitive airport pop out from the dense tree canopy at Belem, she was delighted to spot land. The natives of Belem, after getting over their surprise, afforded her a heroine's welcome.

On March 31, 1964, Joan took off a second time for Natal, which was her jumping point to Africa. She arrived at Natal after more rough weather. Upon her arrival, she couldn't help but notice that the field was filled with armed guards. She was advised to stay at the airport rather than head to a hotel in town: she was the only woman on the base. With both her plane and her room under guard, she ate meals in the officers' quarters of the Brazilian air force and was ultimately grounded there for two days due to political strife. (Note: The 1964 Brazilian coup d'état was a series of events in Brazil from March 31 to April 1 that led to the overthrow of President João Goulart by members of the Brazilian Armed Forces, supported by the U.S. government. This action led to a brutal twenty-year military dictatorship. An April 5, 1964, *The New York Times* article reported that "the Brazilian coup could be regarded as both a powerful shot in the arm for the cause of democratic moderation in Latin America and a bitter defeat for the Communist movements."[71])

The Goulart government fell on the morning of April 2. Subsequently, all domestic and international flights from the airport were cancelled. Because only limited communications were available due to the political conditions, Joan was unable to get a weather report, but she had to leave as quickly as possible. On April 3, she took off from Natal for Dakar, Senegal, in West Africa, with twenty hours of fuel on board. Her next leg was 1,900 miles, with an estimated time of thirteen hours. Battling

---

[70] Ibid, p. 79.

[71] "Brazil Coup Affects Whole Continent; Overthrow of Goulart Is Expected to Bolster the Moderates and Set Back the Communists," Associated Press, *The New York Times*, April 5, 1964, https://www.nytimes.com/1964/04/05/archives/brazil-coup-affects-whole-continent-overthrow-of-goulart-is.html.

violent weather, she arrived in Dakar on April 4. The story now picks up following Joan's departure from Natal.

April 3, 1964: Natal, Brazil, to Dakar, Africa

The Natal to Dakar tape recording reflects my dismal prospects over a destiny shaping up just short of doubtful. I have now been out fourteen hours and have had no contact with the African continent at all. My estimated flight time is one hour over-expended. It took Amelia Earhart thirteen hours twelve minutes to fly the leg from Natal to Dakar and then she hit the African coast, not at Dakar as planned but north of it, and landed at St. Louis, Senegal.

Since leaving Natal, I have spent hours fighting hard against storm clouds, torrential rains, and zero visibility. At one point, it was raining so hard that water starting coming underneath the door, and I thought that the windshield was going to break. I flew as high as 13,000 feet to get above the clouds, but I could never seem to get high enough. So, for two hours, I flew between 400 and 500 feet above the ocean on instruments through a blurry, messy, deluge. Thankfully, after 150 miles of high anxiety and worry, I was able to make it out the other end of that nasty, intertropical front.

I'm about five hours out now from Africa. My thoughts drift to Amelia. I'm flying by magnetic compass only, and am worried that if these winds keep up I might drift too far off course and completely overshoot Dakar. But then, out of nowhere, I pick up Natal radio asking Dakar if anyone has yet heard from me. At first I'm elated, but my joy soon turns back to worry. I yell again and again into the mike, but no one can seem to hear me. Dakar Oceanic Patrol proceeds to contact me on the radio frequencies I am using, but I simply cannot get through to them. Having heard other aircraft throughout this leg attempting to reach both Dakar and Natal with their radios, but with little success, I chalk it up to atmospheric conditions.

It's fifteen hours now since I left Natal. My concern is growing. Once day turns into night, I'm intrigued to see the sea, sky, and horizon all collapse into a single, starless, black void. Suddenly I become aware of how tired I am: my back aches, my head is throbbing, and my leg is cramped. Remembering that I had a couple of cheese sandwiches from back in Natal, I decide to take a bite of one, but quickly toss it aside. It's just too stale.

Flipping on the overhead cabin light, I catch a glimpse of myself. The woman looking back at me is a wild one, all right. I decide to drink some water and munch on a couple of hard candies, full focused on the fact that there's only silence coming from my radio. I switch to a homing device known as VOR. I figure that if I'm within 150 miles of Dakar, I should get an indication on the instrument's needle.

I try again and again to reach Sal approach control. I keep transmitting on emergency frequency 121.5 megacycles, hoping some airlines or ships are in the blackness around me. I know my VHF radio is working. I'm just too far out to receive anyone. Dakar's signal can't be THAT weak. My bearing of Cape Verde is off and erratic. I could fly for hours and in circles and run out of gas. The thought of a night ditching terrifies me. I've never ditched an airplane but I feel I'd have a chance during the daytime. Not at night.

Now, I can't help but think about sharks. I'm as scared of sharks as I am of alligators or piranhas, the last thing I want is to end up down there in the water with them. I've received no altimeter setting in fifteen hours so I don't know if I'm 1,000 feet too high or 1,000 feet too low. I'm flying 11,000 feet indicated, my clearance altitude. Don't know how the winds are. Running out of water. Hungry. No food all day. Keep searching the band on ADF to pick up a station or signal and look it up to see what it could be. In Africa many stations have the same code letter, AF or AG. Three or four stations will have the same identifications and list the names of the cities on paper.

Finally, a signal from Dakar! I discover that I'm on a line only about 40 miles north of the city: I correct course.[72] Dakar approach control advises me to contact Dakar radio on 126.7. Dakar radio wants an explanation

---

[72] Though my intention has always been to maintain the authenticity of Trixie's original manuscript, the April 3 entry was missing considerable information that I felt was necessary to understanding the leg of her journey between Natal and Dakar. In order to bridge this gap and aid the reader in understanding, I have culled Joan's own accounts of this portion of the journey from a variety of other sources, in particular the *Saturday Evening Post* article she wrote in 1965 entitled "I Flew Around the World Alone," and have written this information in what I felt honored Trixie's style and the spirit of the manuscript. A portion of this entry was drawn from this and other sources.

of why I'm late and what has happened. I tell them I was not able to reach them on their published frequencies. I tell them about the high frequency erratic reception due probably to atmospheric conditions. And I tell them about the equator weather I have gone through. They accept this and advise me that they were within one hour of sending out the flying boats. They were going to allow me seventeen hours and then initiate a search from Dakar and also from Natal crisscrossing the Atlantic. I feel a great depth of relief. Glad I didn't try to get into Cape Verde. I can now see lights. The tower clears me to land on runway 2.

Once down, the tower advises me to follow the car leading to the parking area. An African reporter from UPI, a Piper dealer, and some others were on hand to welcome me and say there was food and coffee in the terminal. It was 1:30 a.m., the same as Greenwich Time since Dakar is due south of England. I stay at the hotel on the field where the pilots stay. Nice people. I remained two nights there, as I was exhausted and not quite back to earth so to speak after the ordeal from Natal to Dakar. How dreadful not to have certainty as to where you are.

I sleep till noon the next day when the manager of Air Senegal calls for me to see if I'd like to do some sightseeing. I was intrigued with the natives, tall, dignified, with erect bearing. In the shops I bought some African trinkets, a 14-karat gold cross for 10 dollars, two mahogany figures. The figures are saints, and I selected them because I figured I had something to be thankful for. They would be shipped back to the States for me so I wouldn't be carrying any unnecessary weight.

The evening of April 4 I was invited to a dinner in my honor. I declined as politely as possible. I felt queasy and in need of rest. A fourteen-hour-long flight to Fort-Lamy lay ahead. I was informed that the weather of the moment was good, and I say a silent prayer that it would hold.

April 5, 1964: Across Africa

April 5 I was up at 4:30 a.m. and waiting when the Piper dealer from Air Senegal picks me up to drive me to the plane. Takeoff was at 9:24 a.m. and twenty minutes out I noticed that the generator was overcharging. Battery apparently needed recharging. Gradually the amperage came back to normal so I didn't turn back. Ninety nautical miles out I sighted a beacon. I was under Dakar control at 7,000 feet altitude. Amelia Earhart mentioned in her book the barren terrain in this area. It is like our Southwest. The intertropical front was 10 degrees north. I would have to cross it at about Niamey. Filed a flight plan all the way to Fort-Lamy. Gee (N'guigmi) is under Communist influence right now so I bypassed it, estimating 12 hours en route.

The wide open country was not much to get excited about. I had the high frequency checked and discovered the frequencies I must use were out of calibration. I bisected Africa at its widest section, crossing Kayes, Bamako, Ouagadougou in Upper Volta; Niamey in Niger, Zinder; and Fort-Lamy in Chad. I thought I was rested but soon I felt hungry, tired, and my eyes hurt. There was a haze over the land. It couldn't be smog. Not in Africa. Impossible. I saw a DC-3 out of Dakar and nothing after that. Almost no air traffic. My ground speed was 128 knots. That's 148 miles per hour, and I've had headwinds ever since the flight began.

Dakar's climate was like California's, a welcome relief after Brazil. I was down to 5,000 feet scanning the ground for elephants, lions, zebras. This was their country and natives said I would see them. I didn't see a thing. I could have gone lower, but lower the headwinds were stronger.

Eight hours out of Dakar everything was operating normal. I could have breakfast in Fort-Lamy. But first, I'd land at Niamey to refuel; it had surprises in store. Deep into Africa, it had a long, modern 6,500-foot airstrip, good restaurant, and refueling facilities. Nine hours and 15 minutes after Dakar I touch down on the beautifully lighted field, to be met by the local manager of the Aeroclub and taken to the new three-story terminal. This is a French-speaking city but the control tower operators spoke good English. I was fascinated with Shell Oil Company's refueling procedure in this part of the world. Two 55-gallon drums are hooked to a filtering system in the back of a cart. It is a three-man operation to hand pump the fuel out. It takes 20 minutes to pump 60 gallons. During the laborious procedure a woman aide from the American consulate contacted me and said she was in a hurry to go to a show but offered to take me out to dinner first. I thanked her and declined. I was handed a bill for 250 francs or $2 for landing fee and 3,000 francs or $12 for having the landing strip lights turned on for me when I came in.

I leave Niamey the same night, about 9:41 p.m., and headed for Zinder, making position reports every thirty minutes. Checking the VOR (visual omni range homing device) just before taking off, I found it didn't work. What now? More delay? I explained to Niamey center tower operator and he laughs and said that no wonder I didn't get a signal. The local omni station turned off its electricity to conserve it.

One can't possibly imagine so critical a conservation in the U.S. Total darkness doesn't come till late this time of year. The country was still relatively flat but I did see a few foothills as night approached. As I flew I applied lip chap for my lips, dry from the desert climate, and Murine for my tired eyes. I flew at 7,000 feet since air is more rarified at night than during daytime. I was actually at an altitude equivalent to a higher daytime altitude. So I use some oxygen to keep alert.[73]

Zinder radio beacon was a white spot off to the right. On the other hand, I thought it was more likely a town. With a bit of shock, I realized the white spot on the horizon was the moon. It was a first quarter moon and the first time I had seen the moon come up at 1 a.m.

Night flying is rough on me. At 3:15 a.m., six hours after takeoff, I approached Lake Chad which looked swampy by moonlight. It's just as Amelia Earhart wrote about it. You can hardly tell shoreline from lake. It has no defined shape but perhaps 100 miles long and 80 miles wide. Its shape changes with rainy season. It was a beautiful sight but all I could think of with any enthusiasm was sleep and some coffee. If I put my head down I would sleep, and it was imperative to keep awake and alert so I could make a halfway decent landing at Fort-Lamy.

April 6, 1964: A Sleepless Stretch Across the Sahara

It was 4:30 in the morning when the wheels touched ground and Michael B. Smith from the American embassy approached me. I put only 55 gallons

---

[73] In her article, "I Flew Around the World Alone," Joan writes, "In the African night the stars stood out more vividly than I had ever seen them in my life. They gave me a feeling that I was not just below them, but among them. Below me it was pitch-black, except for tiny red flares that came into view once in awhile from campfires in small Nigerian villages."

in each of the ferry tanks because I expected to reach Khartoum in eight or fewer hours. Black coffee revives me sufficiently to go on without sleep.

At 7:40 a.m. I'm in the air again and recording on tape. I expect to arrive at 3:40 in Khartoum. El Fasher in Sudan would make a nice break in the flight but the landing strip is sand. Not much there. I overfly it. Nor is the wind northeast as predicted. It's north, and I'm getting south of course, so I make a correction. Having a little trouble with the frequency again. I don't know whether my set gets heated up or whether it's the atmosphere. No trouble with it during the night. Wow, I'm 800 miles into the southern part of the Sahara and it's far worse than anything we have in the Southwest of the U.S. No vegetation at all. Just blowing sand. The chart shows a few villages, but I don't see any. Three hours after takeoff I lose radio contact again but I keep trying to pick up Khartoum and other stations.

I have no homing beacon contact ahead yet. I'm tracking outbound on the station behind me. Unless you give advance notice to keep the homing or radio beacons on, the Africans turn them off, to save expenses. Chad will turn theirs on by request. But the flow of traffic over this way is nil. Niger, Senegal, Chad countries all charge landing fees, tax fees, light fees. No wonder traffic is negligible.

Four hours out and still no radio contact. It's extremely rough and turbulent. High noon. Intense heat thermals. Must climb higher. Can't use automatic pilot or can't drink water. Too rough. Very very tired. Only thing worse than going down at sea is going down here.

It's important to remain exactly on course and if you do have to go down, to stay with the plane. This is repetitive of the ocean flight from Natal to Dakar. No contact then. No contact now. Then it was an ocean of water; now it's an ocean of sand, and desolation.

I'm too high to see if there are animals down there, but the heat exceeds 130 degrees so I don't see how anything can live there. I think of the war tales of planes that were down in the Sahara twenty years ago when search and rescue procedures were almost nonexistent. One B-29 that went down during the war wasn't found until 27 years later. The plane was intact, as were the men and their clothing, though they were dead over a quarter of a century.

A man named Smith started out in early 1918 in a plane from Portugal hoping to be the first to fly across the Sahara. Someone had offered him $5,000 for a speculation flight. He took the chance. Forty-two years later he was found. He ran out of gas halfway across the desert and went down. For days he wrote notes, then sat down under the wing of his plane and died.

There's a small town near a hill below, but I see no life whatsoever. Can't reach Khartoum by radio yet again. My heading is 75 degrees. The desert begins to get hazy as darkness again approaches. Khartoum sits on the edge of the convergences of the Blue Nile with the White Nile. After nearly eleven hours of bucking headwinds, I make my approach to Khartoum and finally land at 4:55.[74]

There was no one at Khartoum to meet me. I tried to get in touch with the American Embassy, but the office hours there are until 2:30 p.m. only. I couldn't contact the ambassador. Finally I was able to reach the Associated Press to let them know my landing time.

Shell and Esso people met me there and there was a brief dispute over which fuel I should buy. I was intrigued with the garb worn by the Arabs. They eyed me in my skirt and blouse with equal curiosity. The Khartoum natives have very sharp, pointed features. Different tribes are marked differently with facial designs. Lashes are out in their cheeks at an early age, a form of branding, to identify with their tribe and a certain section of the country.

Some of the people I could identify as belonging to a completely distinct tribe by their lower foreheads, deep sunken eyes, and the beads embedded in their lower foreheads along a horizontal line running from eyebrow to eyebrow.

The people are very black, ebony black. There's much racial unrest, but among themselves. And what large families! I saw women with 14 or 15 children. Wages about $30 a month. Often two women with many children live together. It's impossible that a man can bring home enough sustenance for 30 or 40 mouths with his small salary, but they manage

---

[74] In "I Flew Around the World Alone," Joan wrote, "It was 105 degrees when I landed, and after 36 hours of nonstop flying, I was ready for bed."

to make out. The men wear hats that look hot but are meant to keep out the heat. Women wear three or four different layers of dresses for the same reason. The little ones are strapped on the women's backs and seem contented. I never saw one crying.

When I set out to make a few purchases, the natives stood around wide-eyed. I could take out $20 worth of francs and to their way of thinking squander it on trinkets, when $20 is more than most of them see in a year. Khartoum is different from Fort-Lamy in that Fort-Lamy is a major airport where you can get a plane almost any day of the week to Paris. It becomes more difficult to get a plane from Lamy to someplace in Africa, like Khartoum for instance. French is the language. And the French speak English more understandably than the Brazilians do. At least I think so.

Refueled with Shell at Khartoum airport and met the director of aviation who was in uniform. Aviation personnel have rank and wear stripes on their uniforms accordingly. It's warm here, like our desert in the west. Humidity is low. I heard what sounded like dogs yelping and turn to see Arabs running to catch a BOAC [British Airways Overseas Corporation] jet. Their robes billowed out behind them so that they looked like a band of Ku Kluxers; Sudan Airlines use Viscounts, DC-8s, Comets in their fleet. And in the light airplane class I've seen some French Jodels here, much like our Cessnas.

For hotel accommodations I was offered an adobe hut nearby, but the temperature was 100 degrees and there was no air conditioning. It was imperative that I sleep, so I declined and explain I need a place that was cooler. The airport manager who offered the adobe hut explained it's used by the Arabs who work on the field. Less than three feet off the ground there were apertures for windows, but without glass or screens on them. I could see myself lying awake and hanging onto my purse all night. There were no washing facilities, just nine beds in one big room. I heard of a place in town operated by Americans, and with an air cooling device. How nice it would be to collapse in an air-conditioned room and sleep for a week, waking only for an occasional American hamburger and milkshake. But I must get on with the flight.

The plane and runway will be hot in the morning with a density altitude of about 7,000 feet. I didn't want to stop at Massawa for fuel en route to Aden. Heavily laden and in the heat I would need a lot of runway to get airborne, since the hotter the air, the thinner it is. So I'd have to wait for the weather to cool a bit.

Some American pilots who fly for our government in Sudan, teaching Arabs to be pilots, got in touch with the American embassy. Mr. William Roundtree, U.S. ambassador to Sudan, and an aide, Mr. Beacon, came out to meet me. Mr. Roundtree promptly invited me to his home which is air conditioned; I lost no time in accepting and said I'd be out later in the day.

There was no bottled water in Sudan as I found out in the restaurant where I stopped to eat. The American embassy had said it was a good idea to drink bottled water only, so I asked for it. The natives simply took it out of the tap and put it in a bottle and gave it to me. I asked next for Coke and got some bright red-looking liquid. Ginger beer, they said. One swig and I choked. It was like eating pure ginger or red hot pepper.

Mr. Roundtree's office was a welcome respite. He gathered a couple aides who took me shopping before I had a luncheon with Mrs. Roundtree. Mr. Roundtree came home for the day early. Two-thirty to five is siesta for most workers to avoid the heat of the day. Businesses close too. The most cordial Roundtrees have an 18-year-old daughter finishing high school in Washington, DC. Much traveled, the Roundtrees have been in foreign service 19 years, stationed in Pakistan, Greece, Turkey, Iran, now Sudan. The many mandatory social functions keep them in a social whirl most every night of the week. With five servants, Mrs. Roundtree can retire for several hours in the afternoon but is up at 6 p.m. for the evening's round of activities. It sounds intriguing, but as a routine is tiring and difficult.

The Sudanese were warm, friendly, and seem to like Americans. Some speak good English, most could say hello, and all go out of their way to help me. On the other hand they are an indolent people and take hours to accomplish duties we consider trifling. Most of them sleep out of doors because their huts are too hot. They're all over the streets, at the airport, under planes' wings, on the cool green grass, on benches, at the side of the street. They sleep outdoors until the light wakens them at early dawn and then go inside for what they call their "inside sleep."

The weather bureau and flight office are on the third floor of the airport office building so that I had to step over Arabs curled up against the banisters to get to the weather bureau. They don't like to have their pictures taken, but some posed for me. An Arab woman in a purple lace gown offers to let me take her picture.

At best their ways are not our ways. When Mrs. Roundtree asks a servant to boil some water for my flight and put ice cubes in it, she did just that, put the ice cubes in boiling water. Mrs. Roundtree caught the mistake at the time. She asks also that several sandwiches be prepared for me. They appeared beside my plate along with the usual evening meal. Ambassador Roundtree merely grins and says, "Instructions wrong again."

April 8, 1964: The Arabian Peninsula

At midnight I was back at the field as everything had to be ready for an early takeoff to escape the heat. It took me all of three hours to stir up the proper people to action. Even the controller in the tower was sound asleep, feet on the desk, radio blaring in all directions. Maybe that's why I had difficulty reaching some control towers. Hundreds of Arabs were curled up in every conceivable nook and cranny around the airport. It took me another hour and a half to get the plane refueled and thirty minutes to send two telegrams, one to Jack, one to John Sarver. I had to watch every movement around the plane. The Arabs had dripped gas again. Careless. One was trying to wipe the sand off my windshield with a dry rag, a sure way to ruin the Plexiglass. I knew he meant well, but I was so weary of explaining over and over in each of these countries why you must not wipe dry sand with a dry rag from a windshield.

At 5 a.m. I'm en route to Aden, expecting to arrive there at 9. I planned to get the oil changed, have fuel filters cleaned, pick up AC spark plugs. Since I only carried two extra spark plugs, I ordered others to be shipped to Aden for me. The oil pressure on one engine had been a little erratic. Khartoum people forewarned that Aden was even hotter than Khartoum where summer was just beginning. So it would be admirable to leave Aden for Karachi late in the afternoon after the sun died down. I wonder if it was possible that the prediction for 130 degrees was accurate?

Seven fifteen local time; 5:15 Greenwich time. Now that I was going east from the Greenwich meridian, I'd have to start subtracting from local time, instead of adding, in order to arrive at correct Greenwich.

The ICAO, International Civilian Aviation Organization, decreed here that miles and stateside measures be used again. And just after I got accustomed to using meters. I was passing to the north of the Asmara mountain range [Eritrean Highlands] with its rugged, dry-looking peaks. Some of these Ethiopian mountains go up to 11,000 feet. I had been cleared to fly at 13,500. At Massawa I changed course to Assab, then Aden.

In 1937, Amelia Earhart went into Assab, refueled, and started the long haul to Karachi. She did not land at Aden. Today the sand and short runways at Assab make it inadequate: there are no lighting facilities, the strip is rough, and it's difficult to obtain fuel. I overflew it, taking pictures from the air of both Massawa and Assab.

Ethiopia didn't look radically different from Sudan. But it evoked exotic mental pictures that I remember from school-days geography. It used to be called Abyssinia. And what school child hasn't rolled Addis Ababa around the tongue with visions of giant Nubians and strange cults?

From the air it was just more of Sudan: flat, dry, barren plains. When Jack was stationed in San Diego, our neighbors regaled us with stories of Sudan. They had worked for an oil company in Sudan. From hearing so much about Sudan, Africa's largest country, it seemed to relate more easily to the twentieth century for me. In Khartoum, I longed to linger, to take a ride up the Nile in one of those unique Nile river boats with the unique single sail. And though it was south of Egypt I probably could purchase some of the famed long-fiber Egyptian cotton with its colorful designs.

It was the same story everywhere I landed. I waited so long to see these parts of the world. And now I must be concerned with the airplane, meeting deadlines, and pushing always on. There is no need to be in a rush, except for the restrictions put on my fight by the various foreign countries. I must be in certain areas approximately when expected or go through all the red tape again of securing clearances, permits, even approval to fly over some areas where I didn't intend to land.

So I pushed on, often too busy with the mechanics of flying and navigating to sightsee from the air, and always too conservative of fuel to go scouting around at lower altitudes off course.

I pass Massawa on the west coast of the Red Sea, which is blue, as you'd expect, at 7:55 local time. There was a small airport down there, a single strip about 6,000 feet long with no taxiways. I saw one small twin-engine plane parked. Looked like a housing area just southeast of the airport. It was not much of a town. My flight information manual said that there was gas available there. I drop down to 7,000 feet as headwinds were less strong there. It was more turbulent too, but I preferred to stick out the turbulence in order to fly faster. A low-pressure area is reported south of Aden. The intertropical front was moving in again. This low pressure should give south winds if I stayed on the east side of the low.

Flying down the Ethiopia peninsula now. I looked down on the rugged coast and some swamps. The mountains were 15 miles inland. The country goes sharply from sea level to 10,000 feet. When Amelia Earhart landed at Massawa in the country then known as Abyssinia, she said she could not make a normal glide into the airport but had to snake her way down in big spirals in order to lose thousands of feet of altitude. A normal glide would have put her somewhere in the middle of the Red Sea.

Low stratus condition revealed how, along with fog, because of temperature inversion, the temperature being very hot on the ground and very cool at altitude. The Red Sea is on my left, running all the way down the coast of Ethiopia from Egypt to Aden. Saudi Arabia, and indeed as much as I could see of the United Arab Kingdom, could aptly be named Desolation.

I would be landing at Khormaksar, the airport of Aden, which is operated by the British. It's duty free. The American ambassador at Khartoum said that if I buy anything at all I should do so at Aden, because, like Hong Kong, there's no duty on purchases. I considered getting a camera as I had been taking many color slides, assuming they would all turn out satisfactorily. But since these were opportunities that don't often repeat, I should really have a second camera to take some duplicates to have a pictorial story of my flight.

Ethiopia appeared more colorful, streaked brown, black, and white. The low brush on the desert gives it a brown coloration. Lava flow is black. And the salt beds are white. The latter looked like sand dunes, but Amelia Earhart, too, thought they were sand dunes until she landed and saw that the white was salt in huge piles. The sun, so intense here, draws off great quantities of water each day, and leaves layers of salt for the natives to gather and export. The numerous mountain volcanoes are responsible for the black lava flow.

Flying across Yemen is prohibited so I stayed well south of the island of Perim. When I begin to see Aden ahead of me, I figured about one more hour of flying time. Aden is a heavily traveled major seaport on the sea route to Africa and South America. Saudi Arabia stretched endlessly off to my left, more desolate than anything yet. The charts list Bahr-el-Safi (desert), Rub' al-Khali (desert), and almost no cities: I could count them on the fingers of one hand. At least Ethiopia, besides its change in color, had a few hills, some brush, and the coastal volcanoes to break the monotony of nothingness.

I planned to land about noontime in Aden. As I approached I found it lovely from the air. A 3,000 foot mountain juts out to the southeast. There were many merchant ships in the port. Must be a busy town. The city nestled back against the mountains.

While I close my flight plan on the ground I'm met by Mr. Ramey, director of civil aviation, Mr. Griffin of the American Embassy, and some British RAF [Royal Air Force]. Ground crews started to refuel for me as soon as I was out of the plane. I would stay overnight here and have an oil change for the plane. Well, I had news for my friends at Khartoum. It definitely was not hotter in Aden as they said it would be. The lower temperature was a surprise and a welcome relief after Khartoum. I was mildly amused to see so few Americans here. There are many in Hong Kong and I assumed that at this international port there would be many here also.

Aden Airlines has supplied me with some good mechanics who gave every hopeful indication of knowing something about light aircraft and how to go about the plane with care. While still at the field I phoned to make an appointment in the city at a beauty parlor for a shampoo. By the time Mr. Ramey got me to the Custom Hotel[75] and over to the hairdresser

---

[75] The original manuscript refers to the "Custom" hotel, though this may have been a typo. It's likely that this was the "Crescent Hotel," which was the first hotel to be built in Aden back in 1928. It was refurbished with air conditioning in the 1960s: https://famoushotels.org/hotels/crescent-hotel.

I was late for the appointment and they said they couldn't take me. Surely they could sandwich in a quick shampoo someplace. But it didn't take long to discover that the Aden reception for Americans was, at best, chilly. Natives said the Americans who visit are too cheap, don't spend enough money. Prices of everything in Aden go up when the ships come in. These Aden Arabs with the nasty attitudes are out for every buck they can get, and are so different from the friendly Arabs in Khartoum.

I get into one bargaining session after the other and soon learned the going rates. Consequently, when I was charged 20 shillings for a seven shilling or one dollar ride back to the airport, I knew better and I argued and refused to pay more than the normal rate. I felt the venom of the driver's spate, in a language I didn't understand, and maybe just as well.

Traveler's checks are not accepted in Aden either, except for the banks and major hotels. They wanted cash. Since I was there such a short time, I didn't want a lot of currency as it's frustrating leaving the country with money change in local denominations. The next country may or may not honor it, depending largely on its political attitudes. And I couldn't afford to become a numismatist this trip.

Aden did provide a chance for a good night's sleep if you call 10 p.m. until 2 a.m. a good night's sleep. For me, it was. I got back to the hotel and left a call for 2 in the morning.

= CHAPTER 5 =

FROM ASIA AND INDONESIA, ON TO AUSTRALIA

"Flying by foreigners over Arabia is not welcomed. In the early stages of planning our journey a course was advised eliminating the straight trans-Arabian hop between mid-Africa and Karachi ... as a special precaution we carried a letter written in Arabic, presumably addressed to 'Whom it May Concern' and bespeaking for us those things which should be spoken. At least I think so. We had it translated by two people in New York. One linguist, allegedly familiar with things Arabic, said it would be just too bad or us if such an introduction was presented to the wrong local faction. His council left me a trifle confused. We carried the document anyway, tucked beside me in the cockpit, ready for emergency."[76]

— Amelia Earhart

HAVING BY THIS POINT WRAPPED HER trip across the waistline of the African continent, from the Amazon rainforest through the Sahara desert, Joan prepared to depart from the port city of Aden, Yemen, with plans of next reaching exotic places like Pakistan, India, Thailand, Singapore, Indonesia, and Australia. Chasing daylight while also battling oppressive heat, menacing thunderstorms, aggravating customs

---

[76] Earhart, p. 103.

inspections, and dreaded mechanical issues, Joan carried on with the goal of reaching Amelia's final jumping-off point in Lae, New Guinea.

April 9, 1964: From Yemen to Pakistan

The knock at the door came in the dead of the night, or so it seemed. But my watch agreed with the hotel clerk's. It was 2 a.m. In the lobby I found a friendly BOAC crew also eager to get off the ground before dawn, and we sat out together at the airport and had some hard, black coffee.

By 4:30 a.m. I was taxiing for takeoff when I saw that the fuel pressure was fluctuating quite badly. The right fuel pressure had dropped down to one pound of pressure. I taxied back in. Nothing to do but wait until 6:30 a.m. when the mechanics come on duty. They found that one fuel filter for the cabin tank has collected dirt and must be cleaned. Time consuming. I realized that I won't get off the ground until afternoon.

Takeoff from Aden was uneventful. But since I was carrying full tanks I could climb higher than 800 feet for at least 15 minutes. It was beginning to get hot and cumulus clouds were forming. I could have avoided some of this had I been able to get airborne at 4:30 as planned. The plane wouldn't respond on automatic pilot, which would hold a course heading for me, so I had to hand-fly it for several hours. Finally I managed to climb to 3,000 feet. The weather was good and there was very little turbulence over the water. I lost contact with Aden three hours out, but this is supposed to be normal for high frequency as they don't have much range on their local HF.

I planned to fly to Oman and then change course in Karachi, Pakistan. Oman is an independent sultanate of Arabia. Landing is prohibited anywhere over Saudi Arabia, and special clearance must be obtained to fly over Muscat, Oman. Were I to have an emergency landing, it could be a problem as the British do not have much influence in this sector. My heading was 31 degrees. I put on the Mae West[77] and head out over water.

---

[77] Note: Though it's hard to be entirely sure what is meant by the phrase "put on the Mae West," it may have referred to a shimmy move often performed by Mae West (1893-1980), an American actress who often played the sultry, provocative woman in numerous popular films and plays.

The date was April 9th. 51-Poppa's fuel pressure was fluctuating again. In about 15 minutes darkness came. This would be another night flight. An hour before reaching Karachi I make radio contact and then landed there at 9:23 local time, 16:23 Greenwich. It was pitch dark, but people were waiting for me: Sukira Ali, a flight instructor for the Karachi Aero Club; Adelaide Tinker, the only woman pilot of Sudan and the wife of the American ambassador to Karachi; photographers, and reporters.

The BOAC hotel house on the field was convenient so I stayed at it and BOAC sheltered 51-Poppa in their hangar.

April 10, 1964: From Pakistan to India

In the morning I talked with the women pilots while waiting for customs. Officials put off until morning to put me through their customs inspection. I check weather again and took off—all too soon—but I must be on my way. According to the chart I could save 150 miles by landing not at Delhi as first scheduled but at Ahmedabad, which is also an airport of entry.

Just before I left Karachi April 10th at 9:12 local time, I ran into some friends from Miami who were ferrying a DC-3, Fred Paris and Pete Fernandez. Pete is credited with shooting many MIGs during the Korean War, and for winning the 1955 Bendix cross country speed race in an F-86. The DC-3 took off ten minutes behind me and we landed simultaneously at Ahmedabad.

We got a hot reception: it's 100 degrees on the ground. And I get the usual greeting from customs officials, "Come with me." Such paperwork for such a simple procedure. They questioned my clearance papers which read "Calcutta via Delhi." There's no talking your way out of a situation. No compromises. I sat waiting for three hours.

So different from Amelia Earhart's flight where she reported seldom having customs inspections. Everyone knew her papers were in order and formalities were forgotten, she said, except for the mandatory fumigations of persons and planes. So far the formalities in my case had been attended to with slow, precise, detailed dedication, even though the countries involved knew I couldn't have set out on this flight without all of the proper clearances, permits, papers. The DC-3 crew is waiting here too. They're being held by the senior customs officer because it seemed there was some question about their clearance papers to land at Ahmedabad. I gave the officials a story about landing there because I didn't want to do much night flying. Also mine was an international flight, which stood me in good favor. They decide to let me go.

Amelia Earhart carried with her a to-whom-it-may-concern letter written in Arabic to be presented to nomads, who might or might not be able to read, if by some unlucky chance she were down in the Arab forbidden wastelands. Fortunately she never had to use it. I did some more fast talking about how I would have to be flying mostly at night had I gone as far as north as Delhi, and anyway wasn't Ahmedabad also an airport of entry? They let me take off at 2:30 p.m. their time.

The time I lost on the ground at Ahmedabad I might just as well have been flying to Delhi. I ran into thunderstorms almost immediately. This I dread more than anything. Especially with darkness approaching again. I couldn't see the thunderstorms at night to go around them unless lightning showed the way. And I had no radar to spot them. Calcutta was still 500 nautical miles away. I could have been there one hour before dark had I not been detained by Ahmedabad customs. The country below was dry, mountains were shallow, two thousand feet high at most. The terrain is broken into small farming patches. The monsoons would be breaking in a month or two and then literally could flood the countryside. The monsoons are particularly bad in India, Burma, Thailand and drop up to 200 inches of rainfall.

Further on, in Singapore, Australia, and Indonesia, the monsoons aren't so prominent. But there the intertropical front raised its ugly temper again.

Borneo and New Guinea have been known to have 500 inches of rain in one year. The first big city I identified ahead of me was Nagpur about 375 nautical miles east of Ahmedabad. North of me about the same distance was Agra. That magic name conjured up visions of the Taj Mahal. Having come so far, how I'd love to deviate to see this most beautiful of architectural masterpieces. But Calcutta lay to the east, not north. I changed my course slightly to a heading of 81 degrees. The noise filter in the left magneto was out, causing static on my radio. I could surely get it fixed at Calcutta, as a letter I had received from New Guinea before I left the states assured me there were good maintenance crews in Calcutta.

It's 6:30 local time in India; 13:25 Greenwich. I saw lightning about 45 degrees off to the left. The air direction finder needle tends to point toward thunderstorm disturbances and electrical discharges. Was that another plane approaching? Or was that a star again? The first star came up so suddenly and so brightly each night that usually I was tricked into thinking it was another plane. It was not there and it was there. Maybe tonight it is a plane, since the weather won't permit sight of any stars. I strain my eyes but saw no green or red navigation lights. It must be one lone star shining through the break in thunderclouds. The clouds must be breaking up because 60 degrees to my left I saw a vague Big Dipper. Once the first star was out the others become visible in short order.

At 9:30 local time I got clearance to 9,000 feet by requesting a flight deviation around the thunderstorms. Calcutta was calling me on the approach control frequency. Calcutta asked if, by chance, I had come across any DC-3 pilots. I replied "affirmative" and informed Calcutta that my friends probably still were being held by customs officials in Ahmedabad.

Amelia Earhart related the fun she had in riding a camel in her first day in India. It was such a wild ride that Fred Noonan admonished her to wear a parachute aboard the camel next time. I looked forward to riding a camel, maybe in Calcutta.

The Calcutta tower cleared me to land at Dum Dum Airport. Jim Kiley representing the American consulate, airport manager T.M. Malik, and G.B. Singh, controller of 35 airports in India, met me. Singh is the same man who was in Karachi when Amelia Earhart landed there in 1937. Before I left, I wanted to learn as much as I could about Amelia Earhart from him.

They took me to their air-conditioned VIP room, for which I am grateful. The press interviews begin. The women in the group were wearing saris in lush, striking colors. A news man asked my impressions. I had only just landed but I answered, "I had hoped to see a camel."

"At the airport?" he said and we all laughed. As yet I haven't seen lions or elephants even though I flew low over some parts of Africa. I was about to see animals though, cows which are held sacred in India. Cows everywhere.

I was taken to the Great Eastern Hotel, which is air conditioned. If I seem to dwell longingly on air conditioning, it best reflects my state of discomfort in the extreme heat of these countries. After the crowded, cramped cockpit quarters and the heat and humidity this last week, *air conditioning* were two magic words for me. I used them often and I loved to hear them. Calcutta was somewhat of a surprise in that I pictured it as larger in land size and more cosmopolitan. The population is large enough, exceeds seven million. Great, great poverty. The people huddle, anonymous as ants, by the thousands on the streets where they sleep. The just bunk out wherever they happen to be when night falls. Transient workers cannot afford a roof over their heads. They work for less than 50 cents a day. The American dollar is equivalent to 4.7 rupees. Some workers get less than one rupee a day. Restaurant waiters at the airport get 17 rupees a week.

Natives dress all in white, the men with sheets wrapped around their bottom halves. The sheet is pulled up between the legs giving the appearance of baggy pants. Over this they wear a white jacket. Most of the workers at the field were Indians. Rickshaws popped out of every alley and street, intriguing me. Why did I associate them only with China and Japan? Sometimes children were pulling them, trotting along in the heat like so many agile donkeys. They probably all die at an early age. Traffic was totally uncoordinated. It should legally move on the left side of the street. But here, the right of way is determined by the vehicle having the loudest horn. People hang out the sides of busses.

April 11, 1964: Calcutta

After my night in Calcutta I was driven to the field in one of the wildest thirty minute excursions I've ever had. I'll take my chances in an airplane preferably, anytime. Mile after mile of open air shacks lined the road, shacks with a couple of beds and a cookstove in each. No lamps or electric lighting. Just dogs, cows, and kids running wild. Available meat stalks about in the presence of starvation. But since the predominant religion in Calcutta denies the killing of cows, this means of sustenance is untapped.

At the field I arranged for some minor maintenance on the plane. I had great trust in my AC spark plugs but it makes good sense to have them checked anyway before an over-ocean leg. Mr. Singh is at the field, at my request, has brought some photos to show me of himself and Amelia Earhart in 1937. He reminds me that I was getting closer to Lae, the point from which she took off and never reappeared. I feel a cold chill which I don't manage to shake despite the day's heat. At this point in her route, Amelia Earhart and Fred Noonan had nothing but confidence and assurance of success in their flight. So must I.

Mr. Singh is a pilot himself and won India's first air race. He told me that India was having northwesters and dust storms when Amelia Earhart arrived at Karachi in 1937 and that she had bad weather. In her book she had complained about sand in the Middle East. She also had talked about the monsoon out of Calcutta forcing her to land at Akyab [Sittwe], an unscheduled landing. She took this route in the middle of June, a difficult time of year.

Singh recalled, "Miss Earhart and her companion worked most of the time around her plane at the airport but she did do a little looking around the city of Karachi. Brief sightseeing. Fred Noonan appeared to be a first-class person. They both lived for aviation, always around the plane, looking at charts and discussing their route."

I tell him that Fred Noonan had pioneered the Pan Am routes and was considered one of the best navigators in the world at that time.

Singh continues, "A week or so later we learned she was missing. My God, I was almost certain they'd find them. We were very very sorry that they weren't found. We were hoping against hope. Now it seems like history is repeating itself in another great lady, you. You have much courage to complete the Earhart-Noonan flight."

His words are sweet to my ears, but the personal aspect is somewhat embarrassing too, so I swing the conversation around to him. He got his license in 1929, and in 1932 won the Indian Speedoline trophy, flying an all-wood aircraft with one engine, the Gypsy Moth. He said it cruised at around 85 miles per hour, stalled at 45, and had a maximum speed of 95. I asked if he feels the old thrill in flying has gone. He said, "No, it's more so, in supersonic flight."

I told him about the big air race in our country, the Powder Puff Derby. His Gypsy Moth would have not had sufficient horsepower to qualify for the Derby.

Still reminiscing, Singh says, "India was not divided when Miss Earhart came through. At that time, Karachi was not part of Pakistan, as today, but was part of India. I was born in Punjab."

The dawn arrived and I told Mr. Singh that, fascinating though I found talking with him, I must be on my way. I had a six-hour flight to Bangkok. I climbed into 51-Poppa, was cleared to depart Runway 19 left, and pushed the throttles forward, gaining speed.

I was off at 7 a.m. The friends standing close to the edge of the runway waved goodbye till I was out of sight. There were lots of birds in the air, always a danger to aircraft. The gear was up before I reached 110-mile-per-hour normal climb. My heading was 131 degrees. Amelia Earhart landed at Rangoon but I overflew it, taking pictures, because the Burmese would not give me clearance for more than a twelve-hour period (as I have to pay money just for the privilege of overflying it) to land there. Due to forced delays my clearance has expired. Although I could land there in an emergency. Clouds were scattered over the Bay of Bengal. The heat and humidity were already high and it was only 7 a.m.

I pushed the mic button and told the Dum Dum control tower operators, "Good morning to you. See you again some time." Little did I know how soon.

April 12, 1964: Bound for Bangkok

The temperatures outside were murder on the plane this hot humid April 12. I turned to a course heading of 120 degrees and passed through 6,000 feet altitude at a very slow rate of climb. Strangely enough the temperature, at 6,000 feet, was even hotter than when I took off. They said it would get hotter between 6,000 and 8,000 feet. It reminded me of the Georgia Okefenokee Swamp area back in the States.

There were tiny villages below, which simply float away in the monsoon season. Clouds were forming ahead of me but there were no buildups as yet. I would have 150 miles over water, the Bay of Bengal, before reaching Akyab.

Thirty minutes out of Calcutta, I noticed the left engine had sprung an oil leak. I checked the oil pressure. It was still normal. Because of the six-hour flight ahead I couldn't take the chances.

I would have to turn back and check it out. Very, very aggravating. It could be a broken line. I had to get across the bay before weather set in as it was a bad route.

I radioed Dum Dum. Here I come. And sooner than I had hoped. One reaches a point in bad luck when it's either quit or plod on without dwelling on the cost. Time is money and I was forced to spend time as if I had a monopoly on it.

Several trucks are lined up awaiting my arrival at Dum Dum, my second arrival. I've had more bad breaks on this trip then in my whole flying career. I had them wash down the engine first and then check on it. It had lost about two quarts of oil. There appeared to be a leak at the oil sump bolts. So they tightened up the bolts and it seemed to be all right again.

After one hour on the ground I took off again just before 10 a.m. local time. Akyab was my next checkpoint. Between Rangoon and Bangkok I could expect more cloud buildups. I saw some more oil on the left engine but I decided it was some they didn't wash away since they had washed it by hand. At Bangkok I hoped to find some American mechanics, some who were stationed there to service Pan Am planes. They could check over 51-Poppa thoroughly.

The coast of Burma is all marsh, then a mountain range to break the monotony and then the swampy area again. Rangoon loomed up ahead as my radio reported a low pressure area over Bangkok that assuredly would create weather and reduced visibility. One airplane was somewhere overhead about thirty minutes ahead of me and I hear him calling in, after I had been four hours out of Calcutta. Rangoon looked like a large city, with the airport north of the city. I began tracking into it, careful to avoid the many temples around this restricted area. It's important to look out for them and not overfly them. It's nothing for these people to send up an interceptor plane if they so wish. I wanted to stop there to get the cachets and send stamped envelopes to the U.S. But I was afraid they wouldn't consider that an emergency and my clearance to land at Rangoon had expired.

India must surely hold some dubious record for stalling visitors with red tape, even the natives are friendly and polite, almost apologetic about it all. Amelia Earhart landed at Rangoon in severe weather during the monsoon season and stayed overnight. She had not intended to stop, so her landing there had been nonscheduled. One of the temples below was covered with gold. What a spectacle!

One of Amelia Earhart's favorite songs was "On the Road to Mandalay." Mandalay was up the road several hundred miles, not too far as 51-Poppa flew if I could deviate in that direction. There was a narrow road from Rangoon leading up to it. Amelia Earhart took that road on the ground and found it a cheerful note in the score of flight setbacks.

I have a record of Frank Sinatra singing "On the Road to Mandalay" and back home I play it often. So I sing it now, loud and clear. I decided that Frankie did it better, and quite.

Thoughts of my father were much with me over that remote part of the world. How he loved music, especially Italian tenors. He had some Caruso records that he played frequently.

At 9,500 feet I was estimating Bangkok in about an hour and a half. Weather so far so good this leg. I turned inland over the Burma coast and saw beyond this segment of land, a 100-mile hop over water, the Gulf of Montaban, and beyond, 70 miles or so over land again to the border of Thailand. I was using the turbochargers and making 145 knots (168 mph) ground speed. Whee! Tailwinds for a change. It was extremely hazy over the Burma mountains, making navigation difficult. I requested clearance to fly at 7,500 feet which enabled me to fly below what weather there was. The clouds were 16,000 feet high. Rain showers cut my visibility dramatically and I heard my VOR omni beep for the first time. Knew I was getting close to Bangkok and would have to be careful.

On the charts much of the area in this part of the world is white, indicating uncharted terrain. Boxed notices on the face of the charts advised flyers, "WARNING. Flying is prohibited over the city of Bangkok." And another reads, "WARNING. Low flying is PROHIBITED over the Royal Palace Huahin." So watch out, Joan, for royal palaces. Night flying is also prohibited over Burma except on controlled airways.

"Relief data incomplete" is printed all too frequently on these charts. And in several places it is stipulated, "Numerous buildings that do not cluster into villages are prevalent along the coast and major river valleys of Burma and Thailand."

I landed at Bangkok about 5 p.m. and there grinning stood Fred Paris with the DC-3. He and Pete finally got out of Ahmedabad. This was a religious holiday in Bangkok. I was great at hitting religious or national holidays everywhere I went. Pete and Fred already had arranged for accommodations for me, bless their hearts.

The people were friendly, clean, small-built, and spoke English well. In the flight office I was handed a bill, an atrocious, unbelievable bill. They want a $161 landing fee. They wanted $100 for handling radio fees and my flight plans. I did not fly on instruments but filed a visual flight plan and made six contacts, at the most, en route. I tried to contact the American consulate but was unable to do so. Calcutta had advised me to keep contacts at a minimum en route and I had done so.

Next Borneo Airways hands me a bill for $5 for a plane parking fee, $5 for taking me to terminal building 38, $12 for customs fee (and here I never even saw anyone from customs), $5 for immigration fee, $8 for transportation to town. There are times when pilots appreciate the United States as never before. This was one such time.

I appealed to the airport manager. He was sympathetic but said he couldn't help. Government regulations. I resolved to take this up with the AOPA [Aircraft Owners and Pilots Association] in Washington, DC, when I got back so they could alert other pilots who might be contemplating a flight to the Middle East.

April 13, 1964: Off to Singapore

My first thought was to get out of Bangkok pronto and go on to Singapore. But discretion proved the better part of valor and I decided that sleep, and a normal meal, were more important. Chicken it was, the first time I had chicken in a long time. I checked in early at the hotel and checked out again at 3 a.m. to go to the field, preflight, and get off. The weather bureau was reporting a tropical air mass and thunderstorms over Singapore, but since that's typical for this time of year there was no point in waiting around for better weather. Soekarno, south in Indonesia, was riling up trouble at the moment and Singapore doesn't have much use for its political turmoil in the countries to the south of it. This makes it difficult, weather-wise, for pilots who fly in the area, since unfriendly countries are unlikely to cooperate with one another in providing weather data.

I was in the air at 7:45 flying amber airway number 69 from Bangkok to Singapore. Then I switched over to amber airway 64 changing my course from the east coast of the Malaya Peninsula to the west coast. Sixty-nine parallels the east coast; 64 to the west coast. There were more airports on the west side of the peninsula of Malay, and the chart read, "Numerous kampongs (native settlements) exist in the coastal areas and along principal transportation arteries." I could see the kampongs from the air. They were tucked away in the indentations of the shoreline on the jungle fringes and consisted of wooden huts on stilts.

Singapore sits just north of the equator. Towering cumulus clouds like foreboding sentinels, stood dead ahead of me. Three hours after leaving Bangkok I was approaching Songkhla with the weather worsening as I flew further south. I crossed Kuala Lumpur through rain under a real handicap. The autopilot won't work at all so I had to hand-fly the plane with the charts to follow and instrument navigation. I didn't want to hand fly constantly for 30 or 40 hours over the Pacific as this was too aggravating. It's like a juggler who suddenly finds one hand gone numb.

The old oil leak started again. If not one thing, it was another. The area below was a lush green, indicating plentiful rainfall. It looked like India. Every creek and river had a little town bedded along the banks. I was fascinated with the names . . . Port Sweetenham (British influence obviously), Kampung Telek Merbau, Kampung Siliau, Kampung Telak Merlimau, Bandar Maharani, Batu Pahat. There was Port Dickson. It didn't even belong and for some reason put me in mind of Texas. The strange, fascinating little tongue-twister towns were difficult for me to pronounce.

I usually had not anywhere near the correct pronunciation when calling in position reports. Then the radio operator came back with the right pronunciation I had to double check to see if that was what I really said. The control holds me on visual flight for twenty minutes in instrument weather. Really quite aggravating. Just about over the Singapore omni range I noted the airport runways were wet. So many larger, faster airplanes were coming into the field that the traffic control asked me to hold my position while circling for thirty minutes until they could obtain an instrument landing clearance for me. Comets and jets were all around me. While keeping an eye out to avoid an air collision, I scanned the great seaport of the gateway to India to the Far East. But of course I was more interested in the airport than the seaport and was pleased to find it, for a change, close to the city. Helpful when it comes to transportation. There were boats in the harbor, some surely carrying rubber and coconuts, Singapore's main crops out to the rest of the world.

With my thoughts on temperatures, I discovered my cylinder head temperature gauge had just gone out. One was working, the other wasn't. I knew the temperature in Singapore, or the city of the lion as it is called, hovered around 80 degrees. I had been holding, as requested, flying circles, west of the airport, for twenty minutes. Singapore tower is operated by Singapore Airways. I expected a fee for this. I had hoped it wouldn't be exorbitant as Bangkok's. My holding pattern centered over several small islands just before the airport. Smoke and haze consumed the coastline.

Finally with permission to come down, I landed at Singapore International Airport a few minutes after five. I had scheduled a 2:30 arrival, but coming down the west coast of the peninsula had extended my flight plan. A line squall moving south had paralleled the west coast, but fortunately I flew faster than it was moving. Singapore airport is beautiful, a single, long strip with bright green grass on each side, well-planned buildings, a new control tower, a terminal building to the west of the runway. It stretches belief that this busy jet airport has one runway, however colorful.

The Piper dealer hustled me through customs, and then to the Singapore Flying Club, which boasts 18 planes and 12 instructors. It's the largest such club in this part of the world. Chief mechanic Ruedi Frey, who remembered Amelia Earhart from when she had gone through Singapore, scheduled his men to work on my plane in the morning. There were some minor problems building up due to the buffeting and hard flying 51-Poppa was subjected to. They needed checking. Within two days I'd leave for Jakarta, Java, but the next reliable, fully equipped place to have mechanics work on the plane would be Honolulu. Ruedi Frey drilled the cowling stop, fixed a small oil leak again, cleaned spark plugs, checked the autopilot, and adjusted the cylinder head gauge.

The autopilot is such a help because when fuel dissipates, it causes the weight to shift back and forth creating wing heaviness. William J. Houston from the American consulate drove me to Singapura InterContinental Hotel operated by Pan Am employing Filipinos. The people were courteous, relaying the tales I'd heard of Singapore's corrupt past. It is a regenerated, wealthy city. Traffic was well organized compared with India and some of the other Asian countries I had seen.

I dined on my usual meal of steak, bottled water. I'd stayed away from fresh salads, raw vegetables, fish, dairy products, and so far had defied illness. I was sick once in Cuba from bad water, and sick once in Mexico from a fresh salad. And so learned that if I want to keep well when flying I must give up certain foods in foreign countries.

Our ambassador to Singapore, Sam Gilstrap, and his wife, Mary, took me to dinner, and gracious hosts that they are, they let their guest get to bed at 9 o'clock for a long, late sleep.

While the oil was being changed in the morning and the fuel filters changed, I talked with Ruedi Frey. He was a boy when Amelia Earhart came through Singapore in 1937, but he remembers her. He is deputy chief flying instructor of the Singapore Flying Club numbering 550 members, 100 of them active, and 30 of them student pilots. With five Cessna 172s at their disposal, the club averages 4,000 hours of flying a year.

Ruedi said he took lots of pictures of Amelia Earhart when she stopped to refuel at Singapore. The exact spot she landed was called Kallang Airport and now is just a big park. Ruedi recalled that it was difficult to refuel her plane with its 1,000-gallon capacity inside plus the regular tanks, as refueling had to be done by hand on the sod strip. "There was no radio beacon in those days here, but as I remember, Shell was fairly well organized and got big trucks out to the field to help with the refueling," he said. "I was sixteen years old when she and Noonan came through."

Part of my mission in this world flight is to learn all I can to add to the aviation lore that has grown up around one of the greatest of air pioneers, Amelia Earhart.

April 15, 1964: An Indonesian Adventure

"Tempus is fugiting,"[78] and I must not linger as I'd like to. April 15 at 7:45 I left Singapore for Java. I braced for the equatorial crossing again. I was just south of Singapore and the peculiar position reports in this area all sounded alike to me, all Oriental names that I found difficult to distinguish from one another.

A tight military corridor extends to Jakarta. With much aviation restriction, I was to fly the straight and narrow and not deviate. The first day cover stamps caused extra delay at each stop, but I was committed to them.[79] Storm activity was unusually heavy for the time of year. Low pressure areas were causing two intertropical fronts rather than one. At the equator there was one; another by five degrees south, causing weather south of Java. Singkep radio beacon station was trying to reach me. I relayed a message to another aircraft who passed it along to Singapore. A hundred miles out my VHF (very high frequency) became marginal. It should have better output. And whatever they did to the autopilot, it definitely was not fixed. Still didn't work. One of the UPI newsmen in Singapore offered to wire Mitchell Industries in Mineral Wells, Texas, to find out what might be causing autopilot trouble and to ask for a solution. Possibly the amplifier was out. Autopilot mechanics in that part of the world were at a minimum. I had the fuses changed but it made no difference. It is so tiring to fly hour after hour without an assist from the autopilot.

I ran into the equatorial front 130 miles north of Jakarta. At 7,000 feet I saw lightning slice through the sky. I hated to go through this front.

---

[78] *Tempus fugit* is a Latin phrase that translates to "time flies." The term "Tempus is fugiting" was a famous line delivered by Los Angeles Dodgers sportscaster Vin Scully.

[79] Joan carried 150 Amelia Earhart airmail stamp first-day covers, prepared in 1963 by the Ninety-Nines. They were franked with an additional Earhart stamp. They were cancelled on March 17, 1964, at Oakland, CA. She also carried 1,000 additional covers in a sealed box around the world with her on her flight. A portion of these covers were sold to donors by the Ninety-Nines to collect money for their Amelia Earhart Scholarship Fund. Source: Finkle, Jack. "The Joan Merriam Around-The-World Solo Flight," *The Airpost Journal*, an official publication of the American Air Mail Society, Vol. 36, No. 2, Issue 414, November 1964, p. 31.

I couldn't get much information from Jakarta on high frequency, and I'm too far out to reach them on very high frequency. Low ceilings and high cloud tops assured a lot of bad weather in the front. Trouble began bunching up. The static and turbulence was so bad that I lost all radio for a bit of time. There was nothing but loud screeching noises. It appeared that a loop had gone out. I checked the other and the noise was unbearable. I took 51-Poppa down to 1,500 feet to try to fly under the weather, but the visibility was nil. Jakarta was reporting clouds at 2,000 feet. I was en route through this rugged front for about 100 miles now. Flying on instruments with no radio contact whatsoever is very, very frightening. A thin line of white up ahead raised my hopes that the weather up there was improving.

How blessed to get down at Jakarta and be met by two U.S. Air Force officers, Colonel W.A. Slade and Major T.H. Canady who have been based here a couple of years at Jakarta. Political problems were becoming severe and anti-American signs and slogans abounded. The officers assisted me with filing a flight plan and getting weather reports. At least half a hundred Russian MIG jets were based there and there were many Russian-supplied jeeps about. *Some billboards had a display of American planes being shot out of the sky.*

I hurried to the Indonesian hotel for a bite to eat. Across the street was the British embassy, burned a while back; the black scar had been camouflaged with large boards. There were leftist slogans and pro-Communist insignia everywhere. When I discovered my sunglasses were gone, I hurried back to customs, where I had left them. There were only two men in customs and I had been gone only five minutes, but they denied seeing any glasses. I wasn't in a position to push the matter. This was no place to linger, and I got out as soon as convenient, careful to avoid the highly restricted area a hundred miles along the mainland of Java.

I flew from Jakarta directly to Semarang and then to Surabaya, where I was about six degrees south of the equator between two lines of doldrums. The doldrums are located by measuring the lowest point of low pressure which is 1000.4 millibars at two degrees north and then finding the same millibar pressure south of the equator. Weather was apt to be a little more severe in another 24 hours over the mountainous terrain of this Indonesian string of small islands. As I flew eastward the weather deteriorated and when I landed at Surabaya, a small city on the northern edge of the restricted area, the far eastern edge of Java island, the weather was definitely marginal.

There are a half a million people on this colorful peninsula in the nook of the Madura Strait. As I let down from altitude, judging from the many ships in the bay, I bet this was a fairly active harbor. Rain in the last few days had flooded the inland area. Rice paddies checkered the land and I could see small figures bending over in them. I had the airport runways in sight, but the control tower still wanted me to report over the omni station first so I had to circle around again. I made a 360-degree turn low over the red tile roofs of the city, over the fishing boats in the harbor and what looked like a military fleet. I skimmed over three small planes parked on the landing strip, probably belonging to Garuda Airways. I was down at 5:22 local. I hoped the American embassy could provide a guard for the plane.

Indonesia reputedly has no twilight. There is daytime and then abruptly darkness. At night many radio beacons are shut off to conserve electricity. Not exactly a pilot's paradise. And as the runway was 4,600 feet long, it came at me fast, and I prayed that it was a smooth one since rough runways jostle up the tanks and that's all I needed now.

An American embassy representative and 60 Indonesians form a welcoming committee. My first thought was that perhaps the embassy could provide a guard for the plane. James McHale, director of the Indonesian American Association, Fred Coffey Jr., branch official of the United States Information Agency, and Allan McLean,[80] who, in January, replaced Robert Black as an American diplomat in Indonesia, were on hand to greet me also. I smiled my thanks when they said I could hangar 51-Poppa for the night. There was no such thing, I discovered, as a hotel or public restaurant in Surabaya. I was offered food in the Garuda Airways officers' mess, but accepted McLean's invitation over to his home for dinner. We

---

[80] In a 1964 article from *The New York Times* archives entitled "Operations Closed Down," McLean is referred to as "the United States Consul in the East Java port" of Surabaya.

spoke of the flight and of accepted currency, as the Indonesian currency is now inflating. There are now 330 Indonesian rupees for $1.

April 16, 1964: Approaching Australia

At six in the morning I was back at the field and found some captains and stewardesses also waiting to take off. Breakfast in the officer's mess consisted of rice, sweet and sour pork, fried eggs, tea, and coffee. I settled for hot boiled tea. Maybe I'd prefer rice with chicken or mushroom gravy for breakfast, they asked. "No, thank you. I seldom eat much breakfast," I hedged.

An hour later I was airborne toward Kupang on Timor Island. The Garuda airlines Convairs never go into Kupang because the runway is rough, carved out of coral. I wouldn't land either, unless necessary. Below me sprawled the fabled island of Bali. I could see the volcanoes on this little 70-by-40-mile island. If I had time to kill and this were a pleasure flight, I'd surely drop in, but I had to look ahead 750 nautical miles to Kupang. Winds were easterly—headwinds naturally—10 to 15 knots and even 20 knots (23 mph) closer to Australia. The intertropical front was moving north and I was moving south, and we parted company with no regret.

Each island seemed to have one prominent peak along this chain, making it easy to navigate. "Bali Hai," I hummed to myself. That record was part of my father's collection, with Mary Martin and Ezio Pinza singing *South Pacific* music. All of my father's collection of records disappeared when we moved in 1953. The movers lost a crate of them.

Aboard the plane I had Mr. McLean's homemade cookies and a sandwich from yesterday. And down below me people were dying of malnutrition. I saw not one fat Indonesian. My hunger satisfied, I wished I could just walk into some convenient ladies' powder room. I could see the face now of the fellow at Medina Aircraft in Long Beach when I walked into the office before the world flight and ordered a box of sick sacks, plastic bags. It was the only solution I could think of for long flights when I'd be deprived of bathroom facilities. Male pilots could use rubber tubes, a highly impractical system for women.

"Did you say a whole box, Joan?"

"Right."

"You want ten or fifteen maybe?"

"A whole box."

"You plan to be air sick all the time or something?"

"Just sell me a box full."

"Gal, you don't know how many are in a box. One hundred?"

"That's what I figured. And I don't plan to be air sick at all. Now, may I have them?"

He cocked his head at me and the quizzical look turned slowly to one of understanding. "Sure thing," he said, reddening. "One hundred sick sacks."

They didn't ever really suffice. The cockpit already was space-limited. To maneuver around in it required a finesse that included ability to keep one hand on the controls, and try not to upset the plane's center of gravity so that it would veer suddenly one way or the other. I'm not in the mood for acrobatics.

At Jakarta I had seen naked people bathing in the one canal and using the same canal for bathroom facilities. How impoverished is life in Java. Because of electricity shortage even lights must be off at a certain hour. The electricity is on for a day, then off for a day or twelve-hour period in Surabaya. There were no lights the night I was there. "Just as well," I was told, "lights attract mosquitoes." As an additional mosquito deterrent, I slept under a net. My wandering thoughts had to be caged and disciplined. So, as I flew on to Australia, I picked up the mic and began recording once more.

Flores Island in the Indonesian chain is just north of me. It has three lakes, all close together and different colored. One is red, one blue, and one green. What a color slide it would make! But I won't deviate off course this trip. Considering the hundreds of miles yet ahead before I reach Darwin, I can't fool around taking pictures and wasting fuel.

I'm crossing the Savu Sea 250 miles over open water with a 105 degree heading to Kupang. Weather is good. Ground speed is only 120 knots (138 mph) although I'm truing out at 170 miles per hour in the plane. True air speed almost always is higher than the indicated air speed in the cockpit. Headwinds pass over the island of Timor. Kupang airway is painted on both ends. I try calling their tower but there's no reply.

It may be a part-time operation only. On the other hand, maybe it's siesta time. Elsewhere I've seen control tower operators asleep with feet up on their desks while pilots were trying to reach them by radio. One such negligence in the States and a control tower operator would never see a tower again.

Ahead lies another 480 miles of water across the Timor Sea to Darwin, Australia. My right leg bothers me, cramped. I've been using the turbochargers and have to keep my seat all the way back so I can operate them. My back hurts too. There are several restricted areas out here so that I must remain in radio contact to get through them. Land ahead.

What else but Australia? Great news. I've never relished over-water flying. Possibly because I've never owned a pontoon-equipped airplane. The airplanes in my life number six, one at a time. There was that first Cessna 140 my mother bought, then a BT-13 I bought for $800 in Miami, followed by a clip-wing Piper Cub for acrobatics, another Cessna 140A, a Cessna 172 Skyhawk, and now the twin-engine Piper Apache.

The clip-wing Cub was the result of a sudden infatuation with acrobatic flying. It was 1956 and I attended the Sarasota, Florida air show and watched one of the greatest stunt fliers of all, Bevo (Beverly) Howard. I met him after the show and he let me look over his agile, German-made Buecher Jungmeister acrobatic plane.

I was convinced there could be no more accurate test of flying skill than acrobatics well-coordinated.

My gross-loaded Apache has the aptitude for acrobatics that an elephant has for ballet. A cumbersome possibility. But it is something to think about. I might go back to acrobatic flying with the proper type of aircraft once I finish what Amelia Earhart started, this equator flight.

For the first time I begin feeling I am on the home stretch at last. I have my fifth continent in sight. North America, South America, Africa, Asia, and now Australia coming up. My first real taste of home will be Guam with its U.S. personnel, U.S. currency, and my husband Jack. But Guam is still a long way off.

It's twilight now this 16th of April. About 30 minutes of daylight left. The Australian coast looks mighty good, even though it's ragged and swampy. I ask for a straight-in approach to runway eleven. It is granted. I'm in touch with the tower 100 miles out. I won't dally in Darwin.

I'll try to leave Darwin tomorrow again and continue to New Guinea so I can arrive there before 2 o'clock and get to the post office so I can have the first flight covers stamped. This is a critical stop on the fight. The covers stamped at Lae will mean much, particularly to the Ninety-Nines, as it was Amelia Earhart's final stop. From Lae she took off into eternity. I am eager to arrive there and yet I approach it with some trepidation. I've sent a cable ahead to Crowley Airways at Lae requesting that they try to round up people who were there in Amelia Earhart's time.

Darwin Airport is just ahead. I put my koala bear in the cockpit window so the Australian people here can see him. He has special significance as a good luck piece here in Australia. I only hope no one gives me a real one. I've had enough trouble with the cat back home. I think of Dianna Bixby who tried this flight once and was killed before she got off on the second well-planned attempt. She had named her plane The Southern Cross in deference to her Australian mother.

I like to use the turbochargers as long as possible. I'm starting to drop down at the last moment. The sudden descent doesn't bother my ears. I'm down and told to remain in the aircraft until the authorities come. It's 6:29 Darwin time—two and a half hours difference between Surabaya and here. The fumigation crew should be along shortly. Why else would they ask me to remain in the aircraft? They were fumigating in Amelia Earhart's time too. I can see why they don't want any of the Indonesian mosquitoes here. The city of Darwin itself is on a peninsula several miles from the field. It's not large, maybe 40,000 people. But I shouldn't guess. I'm not so good at estimating populations from the air.

Some places I've seen house 25 and more persons to a shack, and what would accommodate a population of 40,000 in the States will sustain a million plus in Asia or Africa.

Ah, here they come, so I'll sign off this particular session of recording. The fumigation crew will use a spray bomb on the plane inside and out. They sprayed me too. Captain Slade and Ken Davidson who is manager of the Red Flying Kangaroo, Trans Australia Airlines, came up to say hello as soon as the fumigation procedure was finished. They were most helpful in mailing stamped flight covers and some of my recorded tapes.

After the failure of one of her plane's engines, Joan has new engines installed prior to her world flight. (Long Beach, California, Feb. 25, 1964)

Joan Merriam Smith is pictured in the cockpit of her Piper Apache prior to her around-the-world solo flight. Photo dated March 16, 1964.

While grounded in Paramaribo, Suriname, in March 1964, Joan received help from locals with the repair of 51-Poppa's leaky fuel tanks.

In this photo, dated April 21, 1964, Joan stands with the remaining six people in Lae, New Guinea, who witnessed Amelia Earhart's 1937 departure. From left to right: John Cooke, Mr. and Mrs. George Griffiths, Ela Binnell, Joan Merriam Smith, Flora Stewart, and Max Monehan.

On April 22, 1964, Joan arrived at Agana Navy Base in Guam, where she was greeted by a huge crowd of well-wishers, strung with leis, and treated to a special reception.

Upon completion of her world flight on May 12, 1964, Joan is met with plenty of cheer at Oakland Airport.

In late 1964, Joan poses for a picture at Trixie's house with Trixie's children Heidi, Norman, and Patrice, as well as Trixie's husband, Del, and, just to his left, the Schuberts' family friend, Faye Nelson.

A world flight publicity photograph. Photo credit on the back of the image reads, "Lombardi's of Long Beach."

Official U.S. Navy photograph of Joan, dated July 16, 1964.

Trixie waves goodbye to her husband, Del, and their children in advance of the July 1958 Powder Puff Derby. Photo dated May 16, 1958.

Trixie holds the Donald W. Douglas trophy she won in the international literary contest sponsored by the Women's International Association of Aeronautics. To her left is Mrs. Ulysses Grant McQueen, founder of the WIAA. Photo dated May 14, 1953.

From left to right, two FAA representatives and Bill Eytchison stand with Joan Merriam Smith by the wreckage of 51-Poppa.

Burned remains of 51-Poppa following the crash landing in the desert.

= CHAPTER 6 =

INVESTIGATING AMELIA: DEPARTING FROM LAE, ACROSS THE PACIFIC

A T THIS POINT IN THE FLIGHT, progress made on the next leg is best summarized in the form of a press release put out by John Sarver on April 17, 1964. He writes:

> *This is to advise you that Joan Merriam's E.T.A. for Oakland appears to be about Saturday, April 25; as she intends to completely research the Amelia Earhart story in New Guinea and Saipan. I have been in daily contact with Miss Merriam, and the happenings to this girl are unbelievable. If you have been following the daily press stories, you have seen or heard of some of them; but this does not scratch the surface of this uncanny story. It almost seems as if Amelia Earhart was resenting anyone following her path and is throwing the book at Joan to delay and harass.*
>
> *Joan could have twice changed her route and easily been ahead of the other woman for "first woman" honors. Both times she flatly ignored these opportunities and steadfastly clung to the rugged Earhart route. At this writing, half of her radio equipment is*

inoperative, she is hand-flying that plane because of automatic pilot malfunction and her ferry tanks are about as Mickey Mouse as they could be and still hold gas. She has had very little rest because of multiple problems. Even so, she has flown, at this time, further than the other woman's entire trip... but still has a long way to go.[81]

After taking off from Darwin, Australia, on April 17, Joan was quickly met with a fierce storm, forcing her to turn back. In a cable, she reported "world's worst weather over New Guinea!" Amelia Earhart was delayed many days on this leg because of similar weather. The weather is seldom ever fair in this region, but it was something she must fly through.

Attempting a second takeoff from Darwin on April 18, she was pushed again to make a forced landing, this time at Horn Island. By the nineteenth, she attempted to leave Horn Island, but weather stood in the way once again. Later that afternoon when the weather cleared, she was able to make it out and over to Port Moresby, Australia.

On April 20, 1964, Joan departed Port Moresby for Lae, New Guinea, with the intention of completely investigating the Amelia Earhart disappearance. She would soon be faced with additional setbacks and troubles.

April 17, 1964: Extended Stay on Darwin

I stay at Hotel Darwin and, despite getting up at 4:30 in the morning on April 17, do not take off. There is a low loft over the Owen Stanley Range. Weather wide and severe extends for 200 miles. I'll stay another night. Port Moresby can wait a day if it means fighting the intertropical front unnecessarily.

My hosts here are particularly interested in having me see the memorial stone plaque and pillar at Port Darwin, recognizing the first aerial flight from England to Australia on December 10, 1919.

"The pilot's name was Sir Charles Kingsford Smith," they announce proudly, "another Smith like yourself."

---

[81] John Sarver, "Joan Merriam/Around the World Solo Flight," a Sarver & Mitzerman press release, April 17, 1964, courtesy 99s Museum of Women Pilots.

It is here in Darwin that I get my first offer from an air ambulance service to buy 5I-Poppa. I can sell it at a ten-thousand-dollar profit if I wish to abandon the flight at this point and return home on commercial airlines. Once more, in Lae, I'll receive an offer of $31,000, a similar ten-thousand-dollar profit, from Crowley Airways if I'll sell. But I've set out to fly the Earhart equator route and no shortcuts or half measures will suffice. What is ten thousand dollars against ten years of dreaming and planning!

April 18, 1964: To Horn and Thursday Island

Early April 18, Trans Australia Airlines fixes me lunch with hot coffee, serves me breakfast, and makes a tow bar for 5I-Poppa. The other tow bar was lost or stolen somewhere in South America. They've kept my plane in the hangar and watched it for me.

Southeasterly winds with low stratus are moving into Lae. I take off for an estimated eight-hour flight to Thursday Island and vary my altitude between 1,000 and 6,000 feet trying to find a layer with less fog and drizzle. Since middle-eastern Australia has much unexplored terrain, 3,000 feet probably is the lowest safe altitude. I fly in and fly out, above and below the rain, half of the time on instruments.

Darwin radio calls me to warn of increasingly bad weather between here and Port Moresby. I think of going back to Darwin, but that weather is bad too. I must have almost a direct crosswind since I'm correcting 20 degrees to maintain my course. At Cook Island, I break out of the overcast and home in on the Thursday Island navigational sea beacon. Up ahead is the Coral Sea with moist, humid air. Thursday Island is down to one-mile visibility with heavy rain.

There's a Cessna 205 in the air and I talk with him on pilot intercom frequency. He has turned back from Moresby and recommends that I land at Horn Island. Either that or fly down the Cape York Peninsula. I talk next with Starlight One, an Air Force jet taking reconnaissance pictures of weather patterns. I'm on instruments now. I circle the area that should be 25 miles south of an airport, but I can't see any islands. The islands are not as they are pictured on the flight chart. I fly in low circles and through the haze and fog spot two islands. But no landing strip. I finally executed a missed approach procedure and start climbing. At 1,000 feet altitude I glanced out the window and there directly below me was Horn Island airport. I drop the gear and land.

On this estimated eight-hour flight I was two hours overdue, as was my usual luck due to weather. Until we can land on outer-space artificial satellites, and as long as earth is circumscribed by atmosphere, pilots always will be dependent on weather for flights. Weather, of necessity, is an obsession with every pilot.

Since I have been in contact with an aircraft in the area and with Horn Island radio control itself, I am expected. It's a dull, dreary place. No sign of life. Worse yet, no coffee. But I'm glad to be on the ground. Soon I'm joined by Dave Robertson, a Cessna pilot I had talked to in the air. He's Cessna distributor for this whole area. He tells me this is the worst weather he has seen here in about sixteen years.

Eventually, customs, consisting of one inspector, comes over by the boat from Thursday Island to pick us up and carry us back over the choppy sea. There are a couple of hotels in town, he informs me. I look over the Royal. I need a good laugh to lighten this leaden day and the Royal provides it. It's of 1850 vintage and an untidy place. The sheets are what American housewives call tattletale gray and the walls are dirty with an accumulation of years. Well, any port in the storm, and a refuge from the stares of the drunks sitting in a row outside the saloon on the main street.

"Where's the bathroom?" I ask. Considering my sick sack facilities in the air, I feel I'm entitled to the luxury of a bathroom on the ground. The proprietor looks surprised and nods toward the rear of the hotel.

"Out back, about a half a block. You'll see some high ferns, and . . ."

The does it. I'll be happier in the plane, even sitting up all night. Down the muddy street I find another boarding place, the Grant. It has two extra rooms and is somewhat better. I'll stay. Write some letters. No writing paper available they tell me. O.K. then I'll send a wire.

"From here?" they ask in amazement, "not possible." Okay, then a good hot meal. I eat a cold supper. Strangely enough, in this godforsaken place, I find an Australian Broadcasting Corporation representative and spend the early evening talking with him.

April 19, 1964: Horn Island to Port Moresby

In the morning, Dave Robertson meets me at the dock and we take again the dreary, water-sprayed boat trip back to Horn Island. I'd rather give the air a try than stay longer here. We both take off for Port Moresby. The weather thickens and I'm down to 300 feet. Impossible. I turn around and go back to Horn Island where I spend two more hours on the ground checking the weather via Darwin. Darwin is sending out an advisory that a ship northeast of Darwin has reported a cyclone in formation. The low-pressure reading in its center is less than 1,000 millibars. This apparently is what I came through yesterday.

Gradually, the weather improves to the north and I'm off again for another try at 2 p.m. Three hours later I put down at Port Moresby. And there to meet me is Dave Robertson, his wife, and his daughter, Ann, who is learning to fly. A better hotel and a hot meal at last. Trans Australia Airlines is very helpful to me again.

April 20, 1964: Off to Lae, New Guinea

This was the day. April 20 I took off from Port Moresby.[82] Destination: Lae, New Guinea, Amelia Earhart's point of no return. Port Moresby was a pleasant and colorful town but no place to linger. Before I left the airport, Colonel Bayet, who flies a DC-3 out of Moresby, offered to fly ahead and relay the weather to me. At the field they said I was the first person to make a crossing direct from Darwin in more than eight months. That's why there were no pilot reports available on climatic conditions. The weather was so severe on that leg that it took the decals and paint off the side of 51-Poppa.

Yes, it's exciting to look back on these primitive excursions. But the thrill is all in hindsight or aftermath. Such borderline flights themselves as, on the leg from Darwin to Port Moresby, I was so concerned with apprehension mixed with endeavor that there was no time to analyze the flight until later.

---

[82] This is a slight discrepancy. The manuscript refers to the April 20 as being when Joan arrived in Port Moresby, then Lae. A *Ninety-Nine News* article intimates that the twenty-first was the actual arrival date. Here, I have defaulted to the manuscript date.

In a little less than two hours flying time from Port Moresby, I landed at Lae on the north-northeast side of the Owen Stanley mountain range. It was 9 a.m. My en route altitude had been 12,500. Twenty-seven years ago Amelia Earhart landed here at the end of June, leaving on the morning of July 2nd to never again be seen. I felt particularly close to her in Lae and I lost no time in starting to interview the six people whom I asked to come to the field. All six had met Amelia Earhart and were helpful to her in some way when I was in Lae.[83]

April 21, 1964: Extra Day in Lae

My pilot, Laurie, knew the mountains, the crooks around the valleys through them, like he did his own face.[84] A river gushed below us in the canyon winding through the thick trees. It was good to be only a passenger and to have the plane in the hands of such a capable pilot. There was a low cloud scud and to avoid running into the tops of peaks one had to know exactly where the tops were. Suddenly a smoke signal went up and Laurie calmly said, "Ah, here's where we turn to go into the field." It was a hidden, now-you-see-it, now-you-don't field.

Many natives ran out to greet us and some climbed aboard the Cessna to fill their baskets with coffee and the other produce we had along. Laurie sometimes flies up to the mountains three times in one day to bring things

---

[83] In her article "Tribute to a Star," written for the Ninety-Nines' August/September 1964 newsletter, Joan wrote, "That afternoon, we all gathered together for a two-hour lunch. I listened while the others shared their memories of Amelia, and naturally we all speculated what may have happened to her. Twenty-seven years later, the one thing everyone remembered most about Amelia was 'her cheerful spirit and dedication to aviation.' Gratefully I accepted copies of photos that were shared with me, which were taken of Amelia in 1937."

[84] Note: in original manuscript there is no introduction to the section that begins with "My pilot, Laurie." Presumably, Laurie—whether first name or last—was a local pilot who took Joan on a short sightseeing trip while she was visiting Lae. In the *Saturday Evening Post* article, Joan offers some detail about her extra day in Lae: "I stayed an extra day at Lae because I wanted to see more of the place. I took pictures from the air and from the ground of the airport and at the very spot where Amelia's wheels last touched the runway."

to the natives and to reload with their homegrown vegetables which he then flies back to Lae. There is no road. You walk or fly. New Guinea still is the most primitive, unexplored area in the world. Roads run for about 50 miles in each direction along the coast from Port Moresby and then just end.

The land is lush and green. No wonder: 500 inches of rainfall a year. It was a restful respite from my own flying, the flight with Laurie.

I wanted to visit longer in New Guinea. But I had what I came for, the interviews. I tend to believe that Amelia headed for Howland Island, where a landing strip had been made ready especially for her, and was unable to find it. Until there is definite proof that she headed for Saipan and was held prisoner there and perhaps buried there, I will hold to my own Howland Island theory.

I took off from the same runway Amelia Earhart had used, dipping my wings in a farewell and thank you salute to Lae. I climbed on up that morning of April 22 with mixed feelings about myself, Amelia Earhart, and the island-hopping from there on into the States.

April 22, 1964: Naval Escort to Guam

I had some misgivings as a natural reaction to Amelia Earhart's flight from this point onward, but I shoved them to the back of my mind and took up a heading for the island of Guam. I had the best possible motive for getting there as soon as possible. My husband, Jack, was on Guam.

Guam is a Navy base. And I'm a Navy wife. My heart quickened as I got as much speed as possible with the turbo jets, but even then I was too heavy to make good time. The weather proved more cooperative for a change, compensating a bit for the distaste that I have in flying over vast stretches of water.

I concentrated on the instrument panel, on making regular positions reports, and just marking time. There is nothing much to observe over water than water, and the instrument panel in the cockpit. I estimated I was still 400 miles out of Guam. Did Amelia Earhart turn to this area as many think? Was she disoriented? Did she fly this way for a reason? There is no conclusive evidence. All that remains as certainty is that she did not return alive to the United States by any overwater route. Nonetheless I would investigate further, as Fred Goerner had done several times on Saipan, and ask questions there. It was possible that closed-mouthed, reluctant natives would be more willing to talk to another woman following in Amelia Earhart's footsteps, or flight steps.

Guam informed me they were sending an escort for me. But it was still a bit of a shock to see something loom up out of the corner of my eye and turn to find a radar-equipped Constellation flying just off the left wing. Sort of like a minnow being escorted by a whale. He circled around before settling down to fly formation with me. Here was one fish—er, plane—story other pilots would find incredible to believe. He had an indicated airspeed of 150 knots (172 miles per hour) and was doing a magnificent job of slowing down sufficiently to keep pace with me. He told me on pre-arranged frequency to look off to my right and there was a Navy DC-4 joining the formation. Now it was the whale, the minnow, and the marlin.

"Look behind you," I was told. And, heaven help me, there was a jet behind me. It really put me on my mettle, and I tried to fly as precise a course as possible. Now the minnow had the fastest of all sharks on its tail. It was one of the greatest highlights of the entire flight, an occasion my grandchildren will hear told and retold, should I have grandchildren someday.

They flew in such close-knit formation that when my camera case, which I had resting on top of the fuel tank, slipped down, one of the pilots told me on intercom that something that fallen next to me.

"Four nine eight from 51-Poppa," I transmitted, "how long does this formation hold?"

I was informed we would pass over the field, Agana Navy Base, flying parallel to runway six in a flyover just 500 feet off the ground. It was 25 minutes to five and the field was in sight with a huge crowd below. If they thought they were getting a thrill down below, they should have been in my shoes.

With two large aircraft on either side of me and the jet just behind me, we swooped low over the field; then they peeled off and left me alone to proceed with the landing. My heart was pounding hard and even before I rolled to a stop I was looking for the face in the crowd that was Jack's.

Despite all the excitement, something else nagged me. As I put down my gear I realized there was something wrong with the hydraulic system of the gear flaps. I had to return the handle to neutral or there was a hissing sound as of something malfunctioning. I would have to have it checked out, first thing.

The crowd too was in a pique of excitement and people came milling around before I was out of the plane. They strung leis around my neck and shouted words of welcome. What a reception, as only the Navy can do it: Rear Admiral and Mrs. T.A. Christopher—he's commander of U.S. Forces in the Mariana; Governor and Mrs. Manuel F. Guerrero of Guam; Captain Gerald Barlow, who is commander of the Naval Air Station at Agana. All were on hand to meet me. But where was Jack?

Admiral Christopher then handed me a wire from Jack, from the *USS Endurance*. The many unavoidable delays in my flight had brought me to Guam later than my original estimated time of arrival.

Jack was back at sea with his ship. We had missed the mid-Pacific reunion. I was a guest at the home of the Christophers. The Navy VAP-61 worldwide heavy photographic squadron presented me with an album of 8-by-10 pictures showing the otherwise unbelievable flight formation of the big planes with my little one. With dinner came a cake, large and elaborate and decorated with the route of my flight with a nice note: "Compliments and congratulations from the Thomas Bakery." Indeed I was made to feel again the impact of America. Even television after dinner, a novelty after so long without TV. I excused myself and slept early, or tried to. All was so satisfying, except the weather.

While I dozed, the big Constellation typhoon hunters were out getting Navy weather reports. Weather in the area would be marginal, but in the morning I wanted to go to Saipan.

After a sweet ten hours of off-and-on sleep, I wakened to a torrential downpour. The P2V typhoon hunters were still out searching weather, plotting the centers of low-pressure areas. It looked like a new low-pressure threat to the south of Guam. It could move west in a day and I'd have better flying to Wake Island.

April 30, 1964: Saipan Hop

April 24th I took off for Saipan but had to return because the gear wouldn't work properly. I thought the hydraulic power pack probably needed overhauling. The afternoon of April 30th the hydraulic gear plans supposedly were in readiness, so I was off to see Saipan. I took pictures from the air of Rota Island, the first I came upon. It's only one mile-by-one and a half miles in dimension. Tinian Island was next. It was at the Tinian that the first atomic bomb was loaded on the airplane Enola Gay bound for Hiroshima. There was a large, round, concrete circle below. The plane was parked in the circular ramp area to be prepared for its mission and night takeoff. All these islands were infiltrated heavily by the Japanese.

> "I don't really believe the story that Amelia was taken prisoner by the Japanese, but it is now a part of the Earhart legend, and I wanted to see for myself. I talked with some people on Saipan who claim that the Japs shot Noonan and that Amelia died a few weeks later. I saw the place where she was supposedly kept in prison and the spot where she is said to be buried. All fascinating to me, but still only rumor."[85]

May 1, 1964: Second Departure from Guam to Wake Island

Of the eight days I was at Guam, four were utilized in getting the hydraulic gear trouble corrected. Technicians worked for two and a half more days when, on the Saipan hop, I noted a discharge on the amperage. Neither generator was putting out enough power. Radio failure on the Wake Island run, which would be next, could be fatal. As little as ten miles off course without proper radio contact could result in my missing Wake. The ADF

---

[85] As you can see, in this chapter we see another large gap of missing content in the original manuscript. In the April 30 entry, a page is missing from the middle. To bridge this gap and add a bit of clarity for the reader, I've inserted a quote from Joan, which is taken verbatim from her own article, "Tribute to a Star: Flying the A.E. Route," as it appeared on page 82 of the Ninety-Nines' newsletter in the summer of 1964.

requires amperage and it had insufficient electrical output. It had to be corrected no matter what the delay.

May 1, with an honorary medal from Admiral Christopher, a bouquet of roses from the Mariana Lions Club, and other mementos of this momentous stop, I was ready for the next over-ocean leg. The Navy band played as I left Guam this second time.

Mr. Warner of the FAA on Guam warned me that there was a mid-Pacific trough beyond Wake which would make the weather there questionable. I filed a flight plan for an estimated 11 hours en route. I had a long, lonely leg ahead. There was nothing between Guam and Wake but water. All instrument systems seemed A-OK. I took a heading of 74 degrees, correcting for wind.

"Everything is fine up here at 30,000 feet," a Navy plane radioed me.

Great. But who could get up to 30,000 feet in an Apache. Even when not heavily loaded, the service ceiling of my plane is 25,000 feet. There was a C-124 MATS plane in the air and a couple of Navy PVDs. My flight plan called for 9,000 feet altitude. I was almost up to 5,000, climbing heavily and slowly. There were no cirrus or high overcast clouds in sight.

One and a half hours out of Guam, the "up" light on my landing went on. I was dismayed. I should be immune to trouble by now. There was a five-knot decrease in speed. The gear handle was jammed. I realized that somehow the locking mechanism had broken behind the gear handle, jamming inside. I was convinced it was in an intermediate position between the gear well and locked position. If I could just force the gear handle down enough to release it . . .

I called Guam by radio to get in touch with Chief Elbert and ask what I could do mechanically to get the gear handle up. I did not think it was hydraulic trouble but mechanical. I called the aircraft in the air and asked if he had a mechanic aboard. The answer came back: negative.

Within 30 minutes Guam had checked with Chief Elbert, who thought I should return to Guam. He did believe it was hydraulic trouble, and I feared I wouldn't be able to get the gear down to land at Wake and would have to make a belly landing. I reasoned that if I did have to make such a landing, I might as well do it closer to the United States at Wake, then retrace my steps to Guam.

Most of the mechanics in both places are more familiar with large, fast airplanes. When those who are unfamiliar with smaller aircraft begin working on them and get into the various systems, the further they delve, the more they invite trouble.

My indicated airspeed was 100 knots (115 miles per hour). That was slow. If I could just break the jamming hatch. There is a metal plate behind the safety catch, which the handle must bypass to raise or lower the gear. I worked a small screwdriver into the minute opening and pushed the metal tab aside. In my purse I found a piece of tape, and with the other hand taped the metal tab in indent position. Then I passed the handle carefully above this restricted area. I was positive then that the trouble was mechanical and not the hydraulic system.

With all that is required when coming in for a landing, I couldn't be fooling with a tricky gear handle. It would have to get a minimum of attention from me. So I worked it up and down, testing how smoothly I could do it with the tape and screwdriver when the time for letdown arrived.

After four hours of fooling with it, I had the gear up into the well. And the gear light went on again when it should. Guam radio wanted to know how I was doing.

"No emergency," I radioed back and told them I was continuing on to Wake. My true airspeed was 130 knots (149 miles per hour). Despite the report of calm winds, I had headwinds. Wake beacon signal came in strong. It was a good feeling. But I was nine hours out of Guam and could barely pick up the VHF omni signal at Wake. It looked as if I'd make my 11-hour en route time right on the head. Just about dark I could see the Wake runway lights.

I was down at 7:45, Wake time, which is two hours ahead of Guam time. My time en route, 11 hours, 7 minutes. Just seven minutes late.

The FAA literally runs the island at Wake. FAA head man George LaCaille and his bride, Janie, who is an ex-stewardess for Slick Airways, greeted me, along with Father Canice Cartmed, a redheaded priest assigned to the FAA as chaplain. Wake is less than three miles square. Even the commissary is run by the FAA. You might say the FAA inhabits the island—the FAA and the terns. Never had I seen so many birds in one place at one time as there were terns on Wake. They were in the air on my approach just 400 feet off the runway and the control tower warned me, "Beware of birds. Come in high and land low." A nice trick, especially when you were not sure of whether the gear was going down on schedule or not.

There were literally millions of these small, treacherous terns. I was disappointed on landing to learn I had just missed my friend from Nevada, FAA flight inspector Jim Cullerton, who is with the FAA in Honolulu at the moment. Had I arrived a little earlier, our Wake stopovers would have coincided. At any rate, Honolulu lay ahead.

My Apache was under constant surveillance at Wake. But the only menace I could think of was the birds. Janie LaCaille took me to the beach to see them at close range. As we got closer to the shore I saw the beach was carpeted with pebbles, the biggest I had seen, some as large as pullet eggs.

"Look at those rocks. Almost perfectly round. How do they get that way?" I asked Janie.

"Those 'rocks' are tern eggs," she explained. The birds came in by the millions overhead, swooping first in fright and then in anger. Their chirping was so loud and penetrating that Janie and I couldn't hear each other talking. The birds dived closer. Hitchcock made a movie about birds attacking humans. And I had thought it was a joke. No joke. I had the eerie feeling that the birds would kill us if we trampled on their eggs.

"Let's go," I screamed at Janie. I had seen enough of terns.

Wake has 22 inches of rain annually, compared with 75 inches on Guam, and up to 500 inches on New Guinea. I was getting away from the tropical area. It was much drier and more comfortable on Wake than on Guam.

Wake weather was deteriorating again so I waited for it to improve. I sat around hangar flying[86] with Ken Sitton, Pan Am station manager, and Ted Awana, who says that, as a schoolboy, he worked on the airstrip on Howland that Amelia Earhart and Fred Noonan were supposed to land on. Temporary recruiters were being taken for the work. Ted volunteered for 30 days in 1935. (His wife, Lulaine Awana, is president of the Wake Island's Women's Club. She arranged a dinner at Wake in my honor.)

---

[86] "Hangar flying" is pilot talk for when the weather keeps you from flying, so you sit around the hangar waiting and talking about flying.

Search and rescue procedures on these American-occupied Pacific islands are well organized. I talked with Bill Newell and Commander Robert Copes, who fly the big amphibians kept on 24-hour alert for search and rescue.

May 6, 1964: Diversion to Midway Island

It was May 6, and, like a child counting the days before Christmas, I was telling off the days and hours until I would have the USA in sight. I took off at 6:10 from Wake in miserable weather and headed for Honolulu on instruments. I had fairly favorable winds, and by fairly favorable I mean headwinds that weren't quite as strong as on the previous leg. But the headwinds did persist. And the autopilot wasn't working again.

I would be in the air about 15 hours 42 minutes to Honolulu as there were easterly trade winds. That leg was 100 miles shorter than the Honolulu-to-Oakland leg would be. I had just a glimmer of daylight as I took off from Wake. The cloud tops were 10,600 feet high. There is no halfway marker between Wake and Honolulu such as there is between Honolulu and the United States. On that last leg there is Ocean Station November, a ship permanently midway between Hawaii and the U.S.

Midway Island was off to my left, to the north. Johnston Island was to the right. I obtained prior clearance so that if I was not making my anticipated airspeed in the first four hours I could divert into Midway or Johnston.

I could see the strong swell below me. Winds continued easterly. I was approaching 180 degrees longitude, the international dateline. I talked to some MATS pilots in the air. They said the weather was better at 8,000 feet but I was still too heavy to climb that high. And it was cold. My cockpit thermometer showed the outside temperature to be 45 degrees. Whenever I turned the aircraft away from the sun it was cold. My heater wasn't working. But as long as the two engines kept purring along and the tanks of fuel remained intact, I was not about to complain.

There was too much at stake in this flight for me to be upset over some vicious phone calls from Stateside in hopes of injuring or further delaying the completion of the Earhart route. But I was mystified upon learning of the attack I first heard about when I talked with John Sarver from Singapore. He said he had received a phone call from a man back east who also had called Aztec Aviation, Acme Aircraft Company in Torrance, and the Michigan FAA bank that had financed my plane. The gist of the malicious libel was the same in each phone call.

The man said the pilot of the Earhart route, myself, was incompetent and had only 500 hours of flying time. I have over 8,500 hours of flying time logged.[87] He said I did not have the ATR rating, had no instrument experience, and had no money. He came pretty close to being right on the last count. I have very little money. I do have the instrument and ATR rating. The man also claimed to know that I had two stolen engines.

John said he was telling me all this so it wouldn't come as a shock from another source. He also advised that I ignore it.

"Have you done nothing to check out the source of these libelous and malicious calls?" I asked.

He informed me they were the work of a president of a small manufacturer of fuel tanks in the Midwest who had been a partial sponsor of the woman who was flying around the northern hemisphere at the same time I was flying the Earhart equator global route.

The next such attack had been made through Guam where the FAA was wired, before my arrival, that I was flying with fuel tanks that were not airworthy. Pacific region FAA inspector Bob Gale had the tanks inspected at Guam by Mr. Warner and declared they were perfect. I told him about the previous vicious calls and named the source. And received the same advice from him as I had from Long Beach: "Ignore it."

Subsequently, Mr. Gale sent me a letter which read, "We had an inspection, etc."

It was not so easy to ignore the attacks, wondering, at the same time, how much credence had been given the lies back in the States. I would know soon enough.

I checked my speed. It had taken 117 minutes to make 100 nautical miles (115 statute miles). I was concerned about the slow ground speed.

---

[87] The manuscript refers to 8,500 hours of logged flying time. Smithsonian Air and Space Museum records refer to a letter that cites 8,200 hours of logged flying time. Hours may differ simply due to the writings taking place a different times.

By the time I reached 175 degrees east longitude if I didn't have a good bearing, say to Midway, I would know I couldn't make Honolulu with any safe reserve of fuel. In which case I would divert north into Midway. Either that or go beyond the international dateline to Johnston Island.

I had been in the air 45 minutes since leaving Wake and I could hear pilots in the air reporting a heavy overcast over the international dateline. When I no longer could see much ahead I knew I was approaching the overcast. I was running one hour behind time which would make my estimated en route time to Honolulu 18 hours. That was calling it too thin.

I radioed Midway and told them I was diverting to their island. This would take me five hours off course and 200 miles greater distance than if I flew directly. It's a ten-hour flight from Midway to Honolulu. But I saw no safe alternative. I asked for a winds-aloft report. The winds were coming from 320 degrees northerly at 15 knots (17 miles per hour) from 175 degrees east longitude through 180 degrees. I started to descend from 9,000 feet to find better weather below.

At 4:30 local time I was on the ground at Midway and shaking hands with the base commander, U.S. Navy Captain George Washington Davis V. His wife, too, was in the crowd that gathered on learning my flight had been diverted to Midway, most of them government employees. Again the birds proved a hazard on landing. At Wake it was the terns. At Midway, it was gooney birds. I thought the gooney mating season was not until winter. Which showed how little I knew about the mating seasons of birds. They mate—gooney birds, that is—in October and November and only for these two months do they disappear from the island. The gooney bird population on the island is 8,000.

The birds are graceful creatures in the air. But oh, their landings! They'd never pass proficiency tests even in primary flight training. How they ever got their wings I'll never know. Gooneys have heavy breast bones. When they land, they crash land, trying to brake with their leg landing gear but usually nosing over, tail over beak. I found them fascinating and wished I could take a few home to demonstrate to my students how not to land. They have a lovely, graceful, pecking dance. Sort of a two-step. I took many slides of the gooney birds dancing.

Captain Davis said, "That's the mating dance."

"But it's only May," I protested, "and I thought you said . . ."

He shrugged his shoulders. Who knows about gooney birds, for sure?

I planned to take off very early in the morning to beat the birds into the air.

May 6 was the same old story, deteriorating weather. Two high-pressure areas were touching each other and creating a convergence of winds and weather. In short, the pressure systems were fighting each other and didn't want to play umpire up there.

But I made an attempt. I took off and after three hours of flying I was forced to turn back. A Navy plane managed to get through, but at 17,000 feet of altitude and through much turbulence. Overweighted, I couldn't climb that high. It was no good. So back to Midway I went. I crossed my fingers that the gooney birds wouldn't be hovering above the runway to challenge me or to charge into 51-Poppa. I hoped they would go wherever gooney birds go when the weather isn't safe even for birds.

May 7, 1964: Midway to Hawaii

May 7 I tried again, taking off from Midway at 6:10 local time. A Super Constellation had gone up on a test flight to check the weather systems. He reported that the front had dissipated somewhat and was spreading out. I flew for three hours on instruments. Six hours out from Midway I was at 9,000 feet moving into towering cumulus clouds. But I could get around them. The autopilot had been out since Wake and would have to be fixed at Honolulu. Honolulu had omni range. Diamond control began calling me. Cold. Cold. Cold. My coat and heavy socks weren't enough to keep me warm. My feet were numb. I'd get the heater fixed in Hawaii too.

How good Hawaii looked. ONE more hop after Hawaii and I'd be HOME.

May 8, 1964: Heavenly Hawaii

On May 8, my plane touched down in Honolulu. At long last: I was back in the states! Hawaii lived up to its hospitality. I was hung with colored leis, and carnations and roses were pressed into my arms. And there were aviation friends I knew from the past: Sergeant Cartwright, who used to be the Piper dealer at Panama City when I worked for the West Florida

Gas Company; FAA's Bob Gale, who is in charge of an area ranging from Australia to the Far East and to Oakland; FAA public affairs man McCoy; and Bill Snyder. The media had all sorts of questions for me, but the one thing everyone really wanted to know was when I thought I'd be back in Oakland. At that point, I just couldn't say. With both the autopilot and landing gear needing attention, combined with the already unfavorable weather shaping up, the future was just too uncertain.

I was ensconced in suite 16N at the Hilton Hawaiian Village with a view overlooking Diamond Head, and Mary McCloud set about making life as pleasant as she could for me. Never averse to talking about flying, I was pleased to hangar fly part of the evening away with the president of the Hawaiian Aero Club. I would stay in Hawaii until the 25-knot (29 miles per hour) headwinds simmered down.

The airplane needed the rest, and the same could be said for Joan. Foremost on my agenda was a trip out to Diamond Head to see the Amelia Earhart plaque overlooking the beach. In 1935, two years before her attempt at the world equator record, Amelia Earhart was the first to fly solo from Hawaii to California.

Heavenly Hawaii. But I couldn't linger longer in the land of leis with an easy mind while the equator round-the-world flight remained uncompleted. ONLY ONE MORE LEG.

May 10, 1964: The Final Leg

If you can believe it, the words "ONLY ONE MORE LEG" (and in all caps) were the final words written in the original manuscript as it pertained to the tale of the world flight. When I first read the manuscript, I instantly felt deflated to realize that I might never be able to hear what happened to Joan during that last leg home. But luckily, over time, I was able to locate other sources of information that helped to provide some detail about this final leg. Here's a summary of what transpired.

Shortly following takeoff from Honolulu on May 10, Joan noticed that her right engine was beginning to overheat, so she returned the plane for an overnight inspection. Upon taking off again shortly after noon on the eleventh, a few other problems immediately caught her attention. Not only was her landing gear acting up, but she noticed a slight rise in her right engine's fuel consumption. Choosing not to turn back, she spent several hours intently focused on conserving fuel.

Eventually, a bright line emerged across the horizon, catching her eye. It was at that moment when she realized that dawn was breaking along the West Coast. She turned on the radio and could already hear people congratulating her ahead of her arrival. But despite her inclination to celebrate, she had to snap back into reality in order to deal with the crisis at hand—her fuel supply was now exhausted from the right wing tank and beginning to draw fuel from the left side, causing it to overheat. She knew it was only a matter of time before she was going to have to shut it down. So, she did something that she had never done before in her eleven years of flying: she called for an escort to the coast.

Within just a few minutes, Joan was intercepted by a coast guard search-and-rescue plane, approximately one hour outside Oakland. She switched her right engine back on to achieve maximum power at landing, and 51 Poppa touched down successfully at the Oakland airport. Joan arrived to crowds of cheering people. She was handed bouquets of roses, a telegram from Jack, and a wired message of congratulations from President Johnson. To her surprise, she also received an endearing and most timely message from Amelia's sister, Muriel Earhart Morrissey, which read, "In Amelia's name I thank you for your generous gesture in dedicating your flight to her. May this be the first of many triumphs in the air." And with that, her journey was now justifiably complete.

= CHAPTER 7 =

THE FATE OF 51-POPPA

Approximately eight months following the completion of the world flight, Joan was working hard to pay down debt of approximately $17,000 owed on her plane (or $137,000 in 2019 dollars) when she narrowly survived a crash that destroyed her beloved 51-Poppa. It was during this time that she signed a contract with a fellow pilot by the name of Bill Eytchison, who was in need of chartered flight services.

On the morning of Saturday, January 9, 1965, Eytchison was having some work done on his plane in Las Vegas. He contacted Joan to see if she could ferry him back to Long Beach, and she consented. After picking up Eytchison in Las Vegas, on the way back to Long Beach, their plane caught fire, forcing a crash-landing in the Mojave Desert. Following the accident, Eytchison stated that the purpose of the flight had been a business meeting to discuss his and Joan's forthcoming "polar flight," which is explained in the following February 19, 1965, *Los Angeles Times* piece:

> Mrs. Smith planned another adventure to match her 57-day, globe-circling flight around the equator. But this time she would circle the globe by flying on a course that would take her over the north and south poles. William H. Eytchison, 32, of Palos Verdes, who also survived the Jan. 9 crash, said he was to have been co-pilot on

the pole-to-pole flight. "We were going to go in September," Eytchison said. "In fact, I bought the plane two days ago, a twin-engine Aero Commander that cost $117,000. I'm going ahead with the flight." Eytchison said the twin polar flight would have been Mrs. Smith's last. "She planned to give up flying and go into the business end of aviation," he said. "The January crash had left her nervous, and she thought it best to give up flying after the polar attempt."[88]

According to Jack, however, the polar flight with Joan was never going to become a reality. From a February 21, 1965, *Long Beach Independent Press-Telegram* article:

Smith dismissed Eytchison's published statements that Joan had planned a polar world flight in September, with Eytchison along. "There was some talk about it," he said. "But Joan had about written Eytchison off and had almost forgotten about the whole thing. It certainly was not definitely planned."[89]

Jack also vehemently denied that the January 1965 plane crash left Joan feeling nervous. From the same article:

Lt. Cmdr. Marvin (Jack) Smith Jr. angrily denied reports that Joan had lost her flying nerve before her fatal crash near Big Pines Wednesday. Smith reacted indignantly to assertions by William H.

---

[88] Harry Trimborn, "Experts Seek Cause of Women Fliers' Crash," *Los Angeles Times*, February 19, 1965, Main Edition, p. 33.

[89] Lee Craig, "Husband Plans Sky Memorial for L.B. Aviatrix," *Long Beach Independent Press-Telegram*, February 21, 1965, p. 2.

```
Eytchison, 32, who was with Joan when her round-
the-world plane crashed last Jan 9ᵗʰ, that the
incident had left her nervous. "Nonsense," Smith
snapped. "Joan was a professional. To say she lost
her nerve is ridiculous. I flew with her several
times after the Apache went down and she gave no
signs of it."[90]
```

The question looms: was Joan actually serious about completing a pole-to-pole flight with Eytchison? Whose idea was it to complete a pole-to-pole flight, and why? What were Eytchison's intentions? And (if you caught it above) why did Eytchison buy a plane for this particular pole-to-pole flight on the very day that Joan and Trixie fatally crashed on February 17, 1965? Had that been the plan all along? Did he buy the plane before they died that morning or immediately thereafter?

The following text is Joan's own personal account, as told to Trixie, regarding the last flight of 51-Poppa.

The Fate of 51-Poppa

It saddens me less than a year after successful completion of flying the Earhart route to have to add one more chapter to *World Flight* in final tribute to the gallant little plane that made the equator flight possible. It is a "Last Flight" chapter for my twin-engine Apache N3251P, identified simply as 51-Poppa. In a situation that had all the earmarks of a final flight for us both, the broken plane let her life ebb away slowly by fire, giving me time to escape, giving my life continuance.

Together we had gone through some testy times since May 12, 1964, when we touched down at Oakland, California, to cheering throngs and the terminus of our globe-girdling. The Apache had fewer than 300 hours of flying time dating from completion of the equator flight, and I was uncertain what to do with her. Three museums wanted her, but when museums, like the Smithsonian Institution, request historic items, they want them donated. They do not buy them. And I still owed $17,000-plus

---

[90] Ibid.

on the Apache, along with other debts incurred during, and in planning for, the world flight.

I depended on 51-Poppa to help get me out of debt and was taking her to air shows around the country. This, along with some lecture money, was insufficient to support 51-Poppa as her monthly expenses for payments and insurance were about seven hundred dollars.

In a desperate last resort, I advertised 51-Poppa for sale in the Los Angeles metropolitan press. I would take between $25,000 and $27,000 for her. I had several offers for under $25,000 immediately and a call from a woman newspaper reporter who was appalled that I would even consider selling the historic little plane. I explained my problems, and she ran a story about the sentimental attachment I had to my white elephant, coupled with the need to earn money with it or be forced to sell it. Even before the story appeared, in a mood of remorse, I took 51-Poppa off the market again.

Leasing it was a possibility, but the firms interested in leasing wanted it repainted with company colors, which would obliterate the thousands of signatures on the plane.

Any way I turned there was a stumbling block. Then Paul Mantz, co-owner with Frank Tallman of the Movieland of the Air Museum in Orange County, California, and several other donors offered to come up with sufficient money to pay off 51-Poppa and retire her as an item of historic interest, to be preserved for posterity in the museum.

At the same time, as a result of the newspaper story, my mail was flooded with indignant letters from people who offered suggestions and help for saving 51-Poppa. The letters that impressed me most were those from children. It has always been my dual purpose to achieve the equator flight record to fulfill my own great dream and ambition, and through this accomplishment to light the dreams of other young people toward the challenge that a career in aviation affords.

There was never the thought that I would become a heroine in the Earhart and Lindbergh molds of the 20s and 30s; we live in another era. But the letters from so many young people reaffirmed that belief that man always will project himself vicariously through the champions of causes, through men such as John F. Kennedy and John Glenn, regardless of what era we live in. Other mail was advisory—letters from a group of coin collectors suggesting I have a commemorative coin struck and sold, of proposing I get an aviation company to give me a retainer as consultant and pilot until the plane was paid off. And there was one offer, which I decided, after some reflection, to take. The Lorraine Limestone company of Tehachapi, California, wanted me to lease the plane and myself to them on a monthly basis for use in flights to Alaska, Ecuador, and Peru. I signed a six-month contract with a company executive, William Eytchison, liking his attitude that, "You must not lease that plane out without you as pilot. Some part-time flier might take it out and crack it up. And it has too much historic value. After you fly it six months with us and get the bulk of it paid off, it's yours again to put in some museum."

I liked, too, the idea of flying 51-Poppa to Alaska as it was the only state of the 50 in which I hadn't flown her and I wanted her to touch down on Alaska soil before she retired.

January 9th, 1965, a Saturday, I had a phone call from Bill Eytchison in Las Vegas, where he had flown his own plane for some repairs. He asked if I possibly could fly up and ferry him back to Long Beach. I agreed. This clear, sunny, CAVU (ceiling and visibility unlimited) day was destined to be 51-Poppa's last.

When I met Bill at Thunderbird airport he had $1,500 in cash. He handed me some of the bills and said, "Here's $800 toward the Alaskan flight you're making for us. Put it in your purse."

"That's a lot of cash to be responsible for," I said. "You keep it 'til we get back to Long Beach and then give it to me."

"Okay, if you'd rather," Bill laughed. "Let's get going."

I filed a flight plan as usual, noting that the wind was north, five to fifteen knots, and gusting; that visibility was 15 miles-plus; that the altimeter setting was 30.20 inches of barometric pressure. I had 108 gallons of fuel aboard. We took off at noon.

We had been airborne about an hour and were southwest of Daggett on a routine VFR flight when I thought I detected the odor of burning.

"Bill, do you smell anything strange?" I asked, "I have a good sniffer, and there's an odor in here that could be gas fumes. It may be just a gas leak in the electrical gas heater up front here."

I turned it off, but the odor became stronger, more like burning metal. Then we both noticed smoke in the cabin. I shut off all the electrical equipment—the automatic pilot, the radios, the master switch, in that order. Electrical equipment on an airplane operates independently from the power plant, so that the master switch can be shut off without in any way affecting the engines of their output of RPM to the propellers. The heat on the floor became intense, and I shut off the air ducts to the cabin and moved my seat back three notches so that my toes could just reach the rudder control.

We were at 8,500 feet, about 15 to 20 miles southwest of Daggett and holding hands with fiery death. I looked back toward Daggett, 25 miles behind us, and ahead 27 miles to Apple Valley Airport. I could see neither, and it would be impossible to make either one. As it was we'd have to clear two mountain ranges, maintaining our present altitude to do so. We could not risk the time in looking over a landing area, nor was there time to make passes over an area. The first time down would have to be right, landing into the wind as much as possible.

"Let's get down fast." I turned the radios back on, switched to emergency frequency 121.5, and started transmitting.

"Mayday, mayday, this is N3251-Poppa. Fire in forward cabin. Going down. Position, 20 miles southwest of Daggett." I repeated the distress call twice and then shut off the radios and all electrical equipment again on the assumption that something electrical might be triggering the trouble. "There's a flat spot on that mountain ledge."

We headed for it. Antennas on top of the leveled ledge indicated some kind of civilization there. But as we skimmed low, I could see stumps of trees sticking up on the flat surface and hauled the plane away from a landing there, heading it into the valley on the other side. There was a gully, or wash, with a narrow northwest-southeast road running along it. At any moment the fire might break into the cabin. I reached for my fire extinguisher behind the seat. The original one I started out with on the world flight had been stolen in Paramaribo, Surinam, and in Natal I had purchased another. I started to read the directions and then realized the instructions were printed in Portuguese. However, it was the usual CO_2 extinguisher, and I decided it would operate the same as an American-made one.

While we spelled one another flying, we took turns investigating the possible source of trouble. I crawled down under the panel to check for loose wiring, anything that I might correct, but the heat was unbearable.

By now, smoke was burning our eyes, and I asked Bill to get his window open on the left. I was flying right seat and the door was on my side. The Apache has only one door, on the right. The open window created suction but not enough circulation to draw out the smoke. It only stirred it up, so he closed the window.

Bill began calling out the airspeed to me as I said, "I'm going to open the door at 50 feet off the ground to be sure it won't jam or crush closed on landing and prevent us from getting out. She'll shudder and vibrate but have to try to hold her straight."

We had a thermos of coffee along. Bill put it in a bag and said he'd try to get it out with him. I took out the flashlight, a rescue signal mirror, and pliers from the pocket compartment on the side of the door and jammed them into my purse. The thought flashed through my mind that the pliers would be good for something, even though snakes and Gila monsters are not much in evidence in the winter desert.

We were at 4,000 feet now with a northwest heading. Lower and lower until we were rabbit-hopping over the boulder-strewn, roller-coaster territory. The gear was up. A belly landing was the only hope of saving our necks by letting the fuselage take the brunt of the rough terrain. A wheel landing would result in ground looping, sheared-off wings, and worse.

In my heart I knew that there was no hope for 51-Poppa. I could not save my burning plane. With luck, it was just possible the fire would go out once the air ceased rushing through the plane's nose. Flying at 160 miles per hour creates much draft even in a closed area. I felt next to positive that this was curtains for 51-Poppa's occupants as well. I had had my big dream and now it was over, and I was sorry only that I was taking another's life with me.

We checked seat belts. We were almost down before we saw the pipeline road running perpendicular to the wash area we were following. I turned to parallel it. Every 300 feet or so there were piles of anti-soil-erosion dirt and rocks, so we would just have to barrel into these. Nor was the road level. It dipped up and down as much as 60 degrees in some places.

I had my door open now and the plane vibrated wildly. The temptation was great to haul back on the elevator and drop it in, but I continued slow flying, with some throttle, and with one-fourth flaps. Bill called out the speed, "Eighty, seventy-five, seventy, sixty-five, sixty . . ."

Running out of luck quickly. We could anticipate a roll of 300 to 400 feet before coming to a stop, once we hit. One of the intermittent level spots loomed up just ahead and I yelled "NOW!" and dropped the plane in. The props hit first. There was a shrill, shrieking sound like a buzz saw whining through metal, obviously the props cutting through rock. There was an unusual acid smell, in addition to the smell of smoke, and a scraping and grinding as the fuselage just under us ripped open. And then I remember nothing more for a while.

I don't think I was knocked out but was just in a stunned stupor with the enormity of it all. Bill tells me he yelled, "Out. Out." And when I didn't move he shoved me out, and, despite his own bruises and an injured back, he half-dragged, half-carried me away from the burning plane, down a decline near where we had landed.

Bill, who is 6 feet 2 inches tall and weighs over 200 pounds, says that when we hit, the plane plowed straight ahead for three-fourths of the roll and then veered off to the left. In a reflex action, he jumped on the right rudder even though we had no rudder control, skidding as we were on the fuselage itself.

Only five minutes total elapsed from the time we discovered the fire until we were down. But five minutes can be an eternity.

My first clear recollection after the crash landing, my first crash landing in a dozen years of flying, was being on my knees and fighting for breath. I couldn't breathe and did not realize my nose was broken. I was about 50 feet from the plane.

As I turned to look at 51-Poppa, flames shot up from the forward section. I felt a numb horror and pity for the plane, and it was like seeing part of myself burn. There was a succession of muffled explosions and Bill, nearby, called out, "Run. Try to run."

Five-One-Poppa and I had been together for 14 months and 500 hours of flying time, almost four times around the world in distance. I stared, spellbound. My purse, which either fell or was dropped out with me, lay nearby me on the desert. I picked it up and crawled farther away as fast as my bruised body would permit. As I clutched the purse, another horrible thought struck me. The money! The $1,500! I had refused to put it in my purse, asking Bill to hold it until we reached Long Beach so it would be safer.

"The money?" I asked Bill.

"It's in my jacket," he said. "The jacket is still on the seat in the cabin."

Naturally, Bill had taken off his jacket as the heat in the cabin had grown more intense. We looked back at the plane. The seats, or the metal skeletons of them, along with the rest of the cabin, was in smoldering ruin. I remembered the signs that were lying on the seats, too, hand-painted signs to display at airshows, reading, "Please do not autograph the plane—signatures from the World Flight must be preserved. Thank you. Joan Merriam." These would never again be needed.

An hour passed. The flashlight in my purse refused to work. But we had the pliers and the mirror to signal help if other planes should fly over. We limped around until we found the remains of the thermos of coffee in a shattered little pile.

Five-One-Poppa had not exploded on impact. I expected it to, and the thought that it hadn't kept recurring to me, as if the plane had somehow granted me a reprieve even though it would never fly again itself. She had had her moments of grandeur and glory even after the wild flight, as when I was invited to fly in an Air Force F-106B in September. The little Apache flew me to Castle Air Force Base where, after two days of tests and orientation, I was able to fly at speeds of Mach II in the F-106. Then 51-Poppa and I flew the officers and pilot of the F-106 for an orientation ride in 51-Poppa. Our speed? Mach .296.

Where no plane ordinarily was permitted, 51-Poppa was invited. She landed at the Aircraft Owners and Pilots Association meeting at Hollywood, Florida, coming to rest on the Diplomat Golf Club course. We taxied across highways and causeways, through gas stations (without buying gas).

We were down on the desert at 1:30. It was now 2:15 as we stared at the vast, barren land. My flight time had expired and I knew a search would be initiated. All roads led somewhere and we could follow the pipeline road in either direction expecting to find civilization. But pilots are supposed to remain with the aircraft. How long could a person remain without food or water and with lightweight clothing during nighttime temperatures that could dip to 15 degrees Fahrenheit at this 4,000-foot level?

In our still-dazed condition, and because of the peculiar contour of the desert, we did not see the pickup truck ambling across the desert to the plane. It had started away again when Bill spied the departing truck and began shouting. The truck stopped.

John Thiede, miner from Lucerne Valley, had seen smoke rising from the desert floor and assumed that someone had set an old car on fire. He debated whether to go on home 25 miles away and return to his property on the following Tuesday, as were his intentions, or to stop and investigate the fire. Curiosity impelled him toward the fire and, as he got closer, he realized it was not an old car. He took one look at the wreck, wrote down the still-visible numbers of 51-Poppa's identification, added "Burned cockpit. No survivors." He was starting back for Lucerne Valley when he heard Bill yelling. I blanched at the thought of my husband, Jack, receiving that "no survivors" message on his ship in Long Beach harbor. (Jack is commander of the *USS Endurance*.)

Mr. Thiede drove me into Lucerne Valley, where I phoned the FAA to report the accident and called Jack. I learned that Daggett radio had picked up part of my distress message and so were alerted that we were in trouble. From there to the Apple Valley hospital, where I had X-rays. We had sent help immediately to Bill, who remained in the vicinity of the burning plane to warn curiosity seekers who might possibly happen by that the plane's remaining gas tank might yet explode. I estimated there were about 70 gallons of gas left. A sheriff's guard was stationed also at the plane to ward off scavengers.

Jack's first concern, when we were in phone contact, was what damage there had been to me. The X-ray indicated a broken nose. It was the third time my nose had been broken. This time I will need surgery, so I won't have a proboscis that looks like a pugilist's.

As Mr. Thiede had been driving me away from the plane, I glanced back. Though the cabin was totally destroyed by fire, the wings and empennage (tail section) with the bulk of signatures from around the world, remained intact.

I began to hope then. Perhaps some parts of 51-Poppa might yet be salvaged, even though she'd never be airborne again. The right wing was at an upward slant, with the fuel seeping down and feeding the fire. Since fire won't go up a gas line for lack of oxygen inside the line, the fire became a sort of pilot light at the end of the broken fuel line, feeding on the gas as it flowed down. The right main tank was empty but not the right auxiliary tank. The left outboard tank still remained unbroken and was the greatest potential danger of another big explosion.

My beloved Apache, a 1958 model with almost 3,000 hours on her airframe and less than 400 hours on new engines, had received perfect care since I'd acquired her in November 1963. I had spent $35,000 preparing her for the world flight, including new parts, components, extra equipment.

When we left Bill behind on the desert with 51-Poppa, his back began to pain dreadfully. He dipped part of his clothing in gasoline to try to get a fire started with the wood he gathered. But it wouldn't stay lighted. Then he remembered that pioneers used to make a torch or a cow chip and he tried that. It worked. Later X-rays showed that Bill had two chipped bones in his back.

Again I'm in the public limelight but in a manner I would give anything to avoid. Some well-wishers say, "This solves your problem. Insurance will cover the payments on the plane." True, but my plane was my livelihood. Twice I had refused to sell 51-Poppa at $10,000 profit: in April to a charter firm in Australia interested in a turbocharged twin for mountain flying, and again in New Guinea to a flight firm that wanted her for charter.

I shall continue in aviation. It's my life. Already there are some feeble formulations for a new flight in the back of my mind. As for 51-Poppa, Paul Mantz was high bidder for the salvage when the insurance company put up 51-Poppa's remains for bid. A few signatured sections will repose in the Movieland of the Air Museum. It's all that's left as the fire burned into the night and wasn't put out until 12 hours after we were down. For all practical purposes, 51-Poppa is gone. How do you replace the only asset you had?

PART THREE

= CHAPTER 8 =

A PERPLEXING END

ONE IDEA THAT HAS NEVER QUITE sat right for me is that Joan was involved in two back-to-back plane crashes following her world flight, after having completed thousands of successful landings and countless hours of flying. In fact, Smithsonian National Air and Space Museum records show that Joan logged 8,200 hours of flying time with zero accidents, including 5,200 hours in a single-engine, and 3,000 hours in a multi-engine plane.[91] The question soon becomes: were these plane crashes simply an extension of her bad luck? Or could there have been something more?

Upon first reading the manuscript, I was ultimately left with a few nagging questions that I felt pulled to explore:

1. Why were key pages about Joan's time in Lae and Saipan missing from the sequence of pages?
2. Why did Joan pick up Bill Eytchison from Vegas on January 9, 1965, and what really caused the crash of the Piper Apache in the desert?

---

[91] Letter from Betty McNabb, The Ninety-Nines to Mitchell E. Giblo, National Aeronautic Association, March 14, 1965, National Aeronautic Association Archives, Box 86, Folder 11 (Acc. XXXX-0209). Archives Department, National Air and Space Museum, Smithsonian Institution, Washington, DC.

3. Why did a second plane crash happen so soon after the first, and what really caused the crash of the Cessna 182?

I also wondered why Trixie's book was never published if a contract had been lined up with the publisher and the final draft complete. Wouldn't Joan's death have created even more interest in a story about her life and accomplishments? (In April 1965, the Ninety-Nines published in their national newsletter that they hoped to see Joan's book published shortly, mentioning that it might also be released as a movie.) Because my research into these questions took me down various roads, I wanted to explore different scenarios before jumping too quickly to any one conclusion. I began by weighing known intentions against outcomes to arrive at possible explanations.

Did Joan Uncover Something That She Shouldn't Have?

While it could merely be a coincidence, one of the very first things I noticed upon first reading the manuscript was the fact that four key pages from Trixie's original chapter 13 about Joan's time on the ground investigating Amelia Earhart's disappearance were missing. Their absence created a bit of a cliffhanger, because the four pages missing have to do specifically with the time Joan spent in Lae talking to the people who had last seen Amelia, as well as the side trip she took to Saipan. Lae, New Guinea, is, of course, where Amelia last took off before disappearing, and Saipan is the island where she is rumored to have died. I wondered: is it possible that Joan may have discovered something related to her disappearance? Why were these important pages from within the chapter deliberately missing?

The last sentence before the first three missing pages reads, "I lost no time in starting to interview the six people whom I asked to come to the field. All six had met Amelia Earhart and were helpful to her some way when she was in Lae." And then the last sentence before the next missing page reads, "I took pictures from the air of Rota Island, the first I came upon. Tinian Island was next. It was at Tinian that the first atomic bomb was loaded on the airplane Enola Gay bound for Hiroshima . . . all of these islands were heavily infiltrated by the Japanese." Needless to say, we do not know specifically what Joan talked about in Lae or what she saw or did while on the ground in Saipan, as it was intended to be written in the manuscript. While it may be nothing more than a coincidence that those pages went missing, given the nature of Joan's trip to this region, it becomes worthwhile to ask why they would be.

So what is known? We know that Joan had a deep, personal interest in exploring the Amelia Earhart story for herself on the ground in the region where Amelia disappeared. After all, the world flight was conceptualized, planned for, and ultimately executed because of Joan's commitment to fulfilling Amelia's vision. In his April 17, 1964, press release, John Sarver wrote, "This is to advise you that Joan Merriam's E.T.A. for Oakland appears to be about Saturday, April 25; as she intends to completely research the Amelia Earhart story in New Guinea and Saipan."[92] Of course, the words "completely research" indicate a targeted level of commitment versus a passing interest.

In addition to Joan's intentions, we also know something about Trixie's intentions as she kept a variety of research materials related to the book, many of which had to do with the Earhart disappearance. Among those items was a letter written to Trixie's employer from a World War II vet claiming to have met Amelia after she disappeared; a copy of the book *Daughter in the Sky* with a personal note from its author, Paul Briand Jr.; a couple of telegrams sharing thoughts about the disappearance from Amelia's mother; an original 1944 article from *The American Weekly* outlining speculation regarding Amelia's disappearance; a handful of pictures showing Amelia with planes or alongside of President Franklin D. Roosevelt; and even some original Amelia Earhart stamps.

Among all of these items, however, the one that stood out to me the most initially was the letter written by Mr. Roy A. Wykoff Jr., a self-proclaimed war correspondent and former lieutenant colonel in the Tigers Air Force. (During the summer of 1941, retired army Capt. Claire L. Chennault helped form and train a small group of former army, air force, navy, and marine corps pilots to fight the Japanese in China,

---

[92] Sarver, April 17, 1964.

even though the U.S. had not yet officially entered into war with Japan.) The Tigers Air Force, more commonly referred to as the Flying Tigers, was also known as The American Volunteer Group (AVG).[93] In his letter to *General Aviation News* dated July 27, 1963 (see Appendix D), Wykoff claims to have met Amelia Earhart in 1940 while she was living as a missionary in the Marshall Islands. He goes on to explain that she had "escaped from the area of Garapan, Saipan Island, on July 7, 1937 after a Japanese anti-aircraft burst had disabled the airplane in which she was a flier with Noonan." [94] Because at first I didn't know what to make of the letter, having never heard of the Japanese capture theory, I simply ignored it. But once I learned that the Japanese capture theory was a credible hypothesis, I went back and took a closer look. Whether there was any truth to it or not, there was simply something about this letter that stood out to me as uncanny. Its finding would ultimately launch me on a mission to do everything I could to better understand the theories surrounding Amelia's disappearance.

The Era of the Earhart Disappearance

For those unfamiliar with what was happening in the region during the time of Amelia's disappearance on July 2, 1937, allow me to provide a little history. In 1937, the Japanese occupied the majority of the islands in the Pacific while they were also ramping up their control of the region leading up to WWII. On July 7, 1937, the Second Sino-Japanese War began following a minor clash between Chinese and Japanese forces. Incidentally, 1937 was also the same year that soldiers of the Imperial Japanese Army brutally murdered some 200,000 to 300,000 Chinese civilians in the horrific Battle of Nanking.[95] (While there are no official

---

[93] "The Flying Tigers," United States Federal Government, National Archives and Records Administration, accessed January 10, 2019, https://www.archives.gov/exhibits/a_people_at_war/prelude_to_war/flying_tigers.html

[94] Letter from Roy A. Wykoff to *General Aviation News*, July 27, 1963.

[95] Jocko Willink, "The Rape of Nanking," *The Jocko Podcast*, February 1, 2017, http://jockopodcast.com/2017/02/01/60-the-importance-standing-up-against-evil-and-its-heavy-cost-the-rape-of-nanking/.

numbers for the death toll in the Nanking Massacre, these are the rough estimates.)

On September 1, 1939, WWII officially began, and on December 7, 1941, the Japanese made their surprise attack on Hawaii's Pearl Harbor. The U.S. quickly entered into war. During these years, America fought many battles in the Pacific Islands. Chief among these was the Battle of Midway, which marked a turning point for the U.S. once we were able to surprise and defeat the Japanese based on the advancements being made in American codebreaking.[96] Because the U.S. wanted to gain a crucial air base from which it could launch its new long range B-29 bombers directly at Japan's home islands, the island of Saipan would soon become the next major target.

In June of 1944, the U.S. initiated the brutal three-week Battle of Saipan. Arriving at the island in amphibious vehicles, the Americans stormed the beaches, vastly underestimating the number of resistance troops. The Battle of Saipan resulted in more than 3,000 U.S. deaths, with over 13,000 wounded, and the Japanese lost an estimated 27,000 soldiers. When the Americans declared that this battle was over on July 9, 1944, thousands of Saipan's civilians—terrified by Japanese propaganda warning that they would be killed by U.S. troops—leaped to their deaths from the high cliffs at the island's northern end.[97]

In short, at the time of Amelia Earhart's disappearance, tensions were bubbling in the Pacific, international relations were strained, and, frankly, the U.S. would have been interested in any intelligence it may have been able to gather given the impending war, coupled with its interest in acquiring territories from the Japanese. This is why some researchers, like WWII veteran and commissioned air force officer Paul Briand Jr., former secret service agent under the Eisenhower administration Jim Golden, former Federal investigator Les Kinney, and Calvin

---

[96] "The 'Codebreaker' Who Made Midway Victory Possible," *National Public Radio*, December 7, 2011, https://www.npr.org/2011/12/07/143287370/the-codebreaker-who-made-midway-victory-possible.

[97] "Battle of Saipan," History.com, November 17, 2009, https://www.history.com/topics/world-war-ii/battle-of-saipan.

Pitts (best known for his 1981 world flight commemorating Wiley Post), have all speculated that Amelia Earhart was either likely a spy for the U.S. government,[98] on a more casual "white spy" mission,[99] or merely suspected by the Japanese to have been a spy.[100] Any way you look at it, the fact is that if Amelia's plane went down in the Pacific and the Japanese made it to her first, it's safe to assume that they would have perceived any American to be a spy and would likely have treated her as such.

Did the Japanese Capture Earhart and Noonan?

In 1947, there was a significant turning point in American history when the United States government restructured its military and intelligence agencies following World War II. This is, in part, why the public became distrustful of the government's intentions when it came to telling the truth about such things as Amelia Earhart's disappearance or who was really behind the JFK assassination. According to a lecture delivered by Stephen Andrews, the managing editor of the *Journal of American History* and a professor at Indiana University Bloomington who teaches courses in American history:

> If the modern conspiracy theories have a birthday, it is in 1947 with the National Security Act of 1947. It establishes the National Security Council, it established the CIA, it establishes the air force (the air force had prior been part of the army), it established the joint chief of staff, and it creates the CIA. It's charged with "other functions and duties related to intelligence." That is boring

---

[98] Mike Campbell, "Jim Golden and FDR's Amelia Earhart 'Watergate,'" *Amelia Earhart: The Truth at Last* blog, March 13, 2017, https://earharttruth.wordpress.com/2017/03/13/jim-golden-and-fdrs-amelia-earhart-watergate/.

[99] Chris Williamson, "Investigating Earhart: A Conversation with Les Kinney," *Chasing Earhart* podcast, August 12, 2018, https://soundcloud.com/chasingearhart/investigating-earhart-a-conversation-with-les-kinney.

[100] Mike Campbell, "Conclusion of 'Earhart's Disappearing Footprints,'" *Chasing Earhart* podcast, September 21, 2018, https://earharttruth.wordpress.com/2018/09/21/conclusion-of-earharts-disappearing-footprints/.

government speak for secret operations, covert ops ... this is really important because for the first time in American history, the federal government is saying we're going to have a part of the government that is outside of democratic observation. We're going to have a part of the government that will do secret things that we won't tell you about, that we'll spend money on that we will not reveal, and we will do it to protect us because we have to have secrets to protect ourselves.[101]

Coincidentally, it would also be in 1947 that Amelia Earhart's own mother felt it was time to hit the airwaves with a bombshell statement regarding her daughter's disappearance. In one of the telegrams Trixie saved dated May 20, 1947, Mrs. Amy Otis Earhart discusses her daughter's disappearance for the first time publicly after ten years. The telegram reads:

> New York, May 20-(AP)-The Mother of Amelia Earhart, unheard from since she radioed for help while over the Pacific Ocean during a round-the-world flight in 1937, said tonight the missing flier was on a secret government mission and was believed taken prisoner by the Japanese. Mrs. Amy Otis Earhart, 79, of Boston, explained in an interview over radio station WOR that she was discussing publicly for the first time the disappearance of her daughter 10 years ago.
>
> Mrs. Earhart said her daughter was on a government mission "so secret that it was kept even from me," and added: "I think my daughter landed and was taken prisoner of the Japs. I have

---

[101] Stephen Andrews, "Conspiracy Culture in American History," C-Span lecture, July 13, 2018, https://www.c-span.org/video/?448220-1/conspiracy-culture-american-history.

letters, documents, and messages addressed to me that convinced me thoroughly that she landed on land."

The mother said she had made unsuccessful efforts to locate her daughter through the Japanese Consul in Los Angeles. The official with whom she talked originally was "gone," she said, upon her return the following day.

"When I returned there was a strange man there who didn't know the facts, and didn't want to take the matter up," Mrs. Earhart said.

The missing flier's radioed appeal said she was over the Pacific Ocean with no land in sight, and with a dwindling fuel supply in the plane.

United States Navy vessels conducted an intensive search, but no trace was found of Miss Earhart or the plane.

On May 19, 1947, Amelia's mother and her granddaughter flew to New York to make a guest appearance at a Ninety-Nines dinner for approximately one hundred people to celebrate the fifteen-year anniversary of Amelia's historic flight across the Atlantic. Following this event, an interview with WOR radio in New York was arranged for May 20. Mrs. Earhart's announcement led to such headlines springing up across the country as "Amelia Earhart's Mother Declares Famous Flier was on a Secret Government Mission" (*Austin American-Statesman*), "Mother Says Japs Captured Amelia Earhart" (*Tampa Tribune*), and "Amelia Earhart Taken by Japs, Mother Thinks" (*Los Angeles Times*). A search on Newspapers.com confirms that hundreds of news outlets across the country picked up this story on May 21. What did she learn specifically that pushed her to take this information to the airwaves?

In Mike Campbell's book *Amelia Earhart: The Truth at Last*, he concludes—based on years of research—that Amelia Earhart and Fred Noonan were taken as prisoners by the Japanese to the island of Saipan following a crash-landing into the ocean off Barre Island at Mili Atoll in the Marshall Islands, where they were then picked up by the Japanese and taken to the nearby military headquarters at Jaluit Atoll. From there they were flown to Kwajalein and later to Saipan, where they were treated as spies until faced with their deaths.

Among the constant themes of Campbell's book is the idea that the United States government has been covering up this investigation for years. He writes, "If the American public had learned of the abandonment of Amelia Earhart—one of the most admired and beloved figures in our history—on Saipan by President Franklin D. Roosevelt in 1937, his transformation from popular president to national pariah would have been instantaneous, his political future reduced to ashes."[102] In the book, he details the variety of evidence he's collected and catalogued in support of this idea.

Two of the key researchers referenced in Campbell's book are Paul Briand Jr. and Fred Goerner. As mentioned previously, Briand wrote the 1960 book *Daughter in the Sky*, the first to contest the true nature of the Earhart disappearance, and Trixie made a special trip to visit with Briand in 1964. Similarly, Goerner wrote the 1966 book *The Search for Amelia Earhart*, and, as it turns out, Goerner was an acquaintance of both Joan's and Trixie's. On the very day of Joan's world-flight departure, Joan mentioned that Goerner was seated in one of the planes flying alongside her, whose interest in her adventure was "more than just casual." Trixie also wrote in her journal that she had attended an event during September 1964 in rural Hobergs, California, where Fred was the guest speaker. (According to a Ninety-Nines chapter newsletter from 1964, this appears to have been a sectional meeting for the Ninety-Nines, to which Goerner was invited to give a talk.)

---

[102] Mike Campbell, *Amelia Earhart: The Truth at Last: Second Edition*, (Mechanicsburg, PA: Sunbury Press, Inc., 2016), p. 323-323.

Between the years of 1960 and 1965, Goerner famously traveled to Saipan four times to investigate the Amelia Earhart disappearance for himself. Goerner also spent a lot of time poking around high-level government offices for information regarding the Earhart case, having even visited the CIA headquarters, the U.S. military headquarters, the Office of Naval Information, the Pentagon, the U.S. State Department, and many other agencies. At the same time, he was simultaneously broadcasting news about what he was learning through his CBS News radio show. In *The Search for Amelia Earhart*, Goerner speaks of being followed to and from his office, as well as to and from various appointments, by unknown people who were keeping close tabs on him. In short, he was certainly ruffling feathers and making some people nervous at the federal level.

Campbell notes in his book that "only Fred Goerner possessed the courage and insight to write about the establishment's determination to protect the legacy of FDR as the genesis of the government's intransigent policy of deceit in the Earhart case."[103] He goes on to explain the implications:

> The official revelation of FDR's culpability in Amelia Earhart's death would require a rewrite of history, reflect badly on all succeeding administrations for their own failures to reveal the truth, and create chaos among mainstream historians, not to mention our corrupt media. The world wouldn't end, but far too many would be seriously embarrassed and inconvenienced, and maintaining the status quo is a far more attractive option.[104]

Putting the pieces together, it's clear that Joan and Fred were familiar with one another's endeavors. What might the two have discussed leading up to or following Joan's world flight? Could Trixie and Fred have swapped stories about what Joan saw or did in New Guinea and Saipan for use in either of their books? It certainly seems as though

---

[103] Ibid, p. 323.

[104] Ibid, p. 323.

two seasoned journalists, both accustomed to traveling internationally, at a pilots' meeting together at the height of these highly publicized investigations into Amelia's disappearance, would have had plenty to talk about. Could they have talked about their shared interest in Amelia Earhart's disappearance? What about Paul Briand's research, or the NTTU? After all, this would have been just about four months after the completion of Joan's around-the-world flight, less than a few months before Trixie was done with her book, and two years before Fred was to publish his. While it is unknown what Joan, Fred, and/or Trixie may have ever discussed about the Earhart disappearance, the one thing they all had in common was that they were heavily invested in exploring the Earhart incident for themselves.

Joan and Trixie together were well suited to investigate a situation like Amelia's disappearance, and it's completely plausible that they may have uncovered something of interest, even if *they* didn't realize it. Joan was married to a U.S. Navy captain who was stationed in the Pacific. His ship, the *USS Endurance*, traveled to the Far East between January and July 1960 to not only conduct minesweeping exercises at Okinawa, but to provide assistance to the Chinese navy in developing its modern mine warfare techniques.[105] Trixie brought strong investigative reporting skills and potential access to air force insights (her husband coordinated education programs abroad for the air force in Germany; her brother also worked in Turkish intelligence). At the very least, we can establish that Joan and Trixie were eager to collect and capture insights on their own terms, whether or not they had planned to talk about that in the book.

In *Racing to Greet the Sun*, Taylor Phillips shares that in the summer of 1964, Jerrie received a visit from two men at the CIA who showed her pictures of missile bases and asked if she had seen anything like an American base on her trip. Jerrie suspected that photos taken during

---

[105] "USS ENDURANCE (MSO-435) – an aggressive class minesweeper," Hull-number.com, accessed February 18, 2019, http://www.hullnumber.com/MSO-435.

her world flight were rerouted to the CIA.[106] Could the CIA also have had an interest in the knowledge Joan obtained while traveling solo around the world? After all, she had mentioned taking pictures of Tinian Island, where the first atomic bomb was loaded onto a plane bound for Hiroshima. Did Joan actually see or take pictures of anything she shouldn't have? Was Trixie planning to write about it? Was there someone who didn't want to see the book published based on information that the two women may have collected? Perhaps the missing manuscript pages were simply lost after Trixie died, or maybe Trixie had pulled them in order to rewrite them. Or, could someone along the way feasibly have borrowed them because they were interesting enough to keep?

As for that letter Trixie saved from Wykoff about having met Amelia in the Marshall Islands? Multiple sources state that the letter is completely bogus, and I was never able to find evidence that such a person existed in the Flying Tigers Air Force. But, according to the *Quad-City Times*, a Mr. Roy A. Wykoff—whose address matched that of the letter writer—did exist. He died at age 48 on April 27, 1965, nearly two years after the date on that mysterious letter, at the Veterans Hospital in Iowa City, Iowa. His obituary states that although he never married, he was a noted WWII veteran.[107] Were there any kernels of truth to his story, or was it simply a tall tale from a beleaguered veteran, looking for an opportunity to raise his stature or validate his self-importance in the public's eye following a brutal and unfulfilling wartime experience? Regardless, the fact remains that Trixie kept the letter and found it interesting, which deserves some merit given her background as a foreign news correspondent and investigative journalist.

Was the Crash of 51-Poppa Truly Accidental?

If simple bad luck was to blame in both of Joan's plane crashes, that's one thing. But can two plane crashes less than five weeks apart really be pinned to bad luck? While at first I hadn't considered the idea that

---

[106] Phillips, epilogue.

[107] "Roy A. Wykoff," Official Notices/Obituaries, *Quad-City Times* (Davenport, IA), July 24, 1965, p. 13.

Joan's crashes could have been anything more than two unfortunate accidents, upon closer inspection I certainly was left wondering.

A good place to start with is Joan's husband, Jack Smith. In a video interview on YouTube dated December 2016, Smith reflects on Joan's accomplishment of flying around the world. He also shares his account of what he remembers about both crashes. In reference to the first crash, he begins:

> Lo and behold, she was flying somebody to Las Vegas. Those airplanes had a gasoline heater in the nose that heated the airplane at altitude, and that heater caught fire. The smoke came roaring through the airplane and filling the cabin, and they opened the doors, and they were choking and everything else. They were in the mountains, somewhere between Las Vegas and Long Beach, and they elected to put it down on the pipeline maintenance road, wheels up. So they went down wheels up and successfully landed, practically destroying the airplane. Her passenger, who was some young millionaire, actually had to pull her out of the airplane. She had gotten a bump on her head, she was dazed, and he pulled her out. We lost the historic airplane at that time.
>
> So later on, not long after that, one of the aviation writers, who was also a young female pilot, Trixie Ann Schubert, was doing a story of her flight around the world. The short story actually got published in the *Saturday Evening Post*. Trixie and Joan were over the mountains east over San Diego, and all of a sudden, there was extreme turbulence; you'll find that flying around the mountains. The upper, attached fitting of one of the wings broke, and the wing folded alongside of the airplane, and the airplane hoggered in and hit on the 7,000-foot level above the sea and destroyed the airplane and both girls perished."[108]

---

[108] Jack Smith, video interview series courtesy of Vic Cambell (Me3TV YouTube channel), December 2016, https://www.youtube.com/watch?v=R9X-8hJ1WuDQ.

Jack's recollection of events provides a great baseline for exploring further questions. From what he remembers, he pins the cause of the first crash on a gasoline heater catching fire, and the cause of the second crash to turbulence that caused a structural failure.

51-Poppa Under Fire

Joan certainly had her fair share of troubles with 51-Poppa. In fact, three weeks prior to her around-the-world flight, Piper Aircraft's Jake Miller was quoted as saying that he didn't want to sponsor her flight because of the plane's poor quality. "Hers was an old airplane, of no particular interest to us. It had been converted and re-converted so much that we felt she was going to have a lot of problems. We didn't know anything about her fuel system—who had designed it—or her turbochargers."[109]

During the flight, of course, there were emergency landings, major setbacks in Brazil and Guam, as well as a couple of close calls due to the pressures being put on the plane from the stress of a world flight. But if an around-the-world flight were going to be easy, wouldn't someone else have completed it already? Joan explained why she chose 51-Poppa for her around-the-world flight:

> I chose a 1958 Apache, equipped it with turbochargers to improve its performance remarkably . . . I felt I had a good chance of flying this ship at and below the equator to duplicate the Amelia Earhart route. Many advised against such a route for I would have a constant array of weather and headwinds problems throughout. Everyone said I could not do it in an Apache, and advised everything from a twin-engine Bonanza to a B-25! In as much as I could only afford an

---

[109] Gilbert, p. 81.

```
Apache and this sounded like a challenge, I was
ready to go![110]
```

In short, even though Joan was somewhat limited by what she could afford, she felt comfortable with her plane, and she was ready to commit to the challenge. Twice previously she had located planes for purchase, but both times the plans had fallen through financially.[111] The time was now or never.

Following the world flight, Joan owed $17,000 on her plane, and estimated the value of publicity to be more than $300,000 to Piper alone. Had she completed the flight ahead of Jerrie, positive publicity—e.g. paid speaking engagements, etc.—would certainly have helped to eliminate her debt. But this was not the reality, so she had to put her head down and get to work.

Trixie first heard about Joan's plane crash with Bill Eytchison on the radio, while doing dishes at two a.m. In a personal journal entry dated Saturday, January 10, 1965, she writes:

As I was doing the dishes at 2 a.m. today, I heard night owl radio program announce that Joan Merriam had crash landed her Apache in the desert after a fire in mid-air. Read the story in the L.A. Times after mass. She has broken nose, abrasions, etc., but is alive. I tried to reach her by phone—unsuccessfully—so I sent her a special delivery. Also wrote a note to Roy Bailey (who checked me out in the 182 race plane last year for the Powder Puff) and who was hospitalized after forced landing when engine quit over Long Beach State College. I wait for something to happen to a third pilot friend, as such news, usually and incredibly enough, breaks in threes.[112]

---

[110] Joan Merriam Smith, "Tribute to a Star: Flying the A.E. Route," *Ninety-Nine News*, August/September 1964, p. 12, https://www.ninety-nines.org/pdf/newsmagazine/19640809.pdf.

[111] McNabb.

[112] Trixie Schubert's personal journal, January 10, 1965.

But then on Monday, January 17, 1965, she says that her worries were eased: "Marina called me to read me a story about Iris Critchell who made an emergency landing when a buzzard broke through windshield of her plane and landed on her lap in the cockpit. That's the third."[113]

According to the official NTSB crash report for N3251P, the cause of the crash on January 9 was due to "fire or explosion: in flight."[114] Joan described the sequence of events as beginning with a burning metal smell, followed by smoke, and, finally, fire. I wondered: what would actually cause a gasoline heater to suddenly catch fire in the nose of a plane? Was the plane just too old, tired of facing repairs and modifications? Or had faulty wiring been the cause? Could the heater have been sabotaged? Who was the last person to perform work or maintenance on Joan's plane? She did mention that the heating and electrical systems had been carefully checked just two days prior to flight in preparation for a forthcoming Alaskan flight.[115] Did she land and immediately take off from Vegas, or did she leave the plane unattended for some time before flying back to Long Beach? Needless to say, the fire in the cabin that caused 51-Poppa to crash still isn't fully understood, and the full accident report has long since been destroyed.

Bill Eytchison: The Millionaire

In reading about the crash of 51-Poppa in the context of history, I immediately wanted to know more about Bill Eytchison. Why did he need a "ride home" from Vegas? Three things we know about Eytchison from his flight with Joan are that he carried around plenty of cash, he was involved in a mining operation, and his business activities required regular air travel.

Prior to the final flight of 51-Poppa, Joan had been contracted by Eytchison to be a private pilot for chartered flight services. As such,

---

[113] Trixie Schubert's personal journal, January 17, 1965.

[114] "NTSB Identification: LAX66F0066," National Transportation Safety Board, File 3-2184, https://www.ntsb.gov/GILS/_layouts/ntsb.aviation/brief.aspx?ev_id=76822&key=0.

[115] Joan Merriam Smith, "The Fiery End of '51 Pops,'" *The AOPA Pilot,* March 1966.

Joan had an arrangement with him to provide air transport when needed. On the day Joan picked him up in Vegas, he was carrying $1,500 in cash. This fact stood out to Joan, who actually remarked "that's a lot of cash to be responsible for." Approximately $800 of that money, however, was designated for a future flight that Joan would be making to Alaska on his behalf. The contract for air services was ultimately a way for Joan to help pay off her plane.

Among the interesting stories I came across regarding Eytchison was one from March 1953 in which he and his father, Arthur (also a pilot), famously used the radio to help talk another pilot—who was flying over Los Angeles and threatening to crash his plane—out of committing suicide. Thirty-year-old former air force pilot Roland Clarence Sperry had then been an employee of Eytchison's.

A *Los Angeles Times* article from March 28, 1953, mentions that Eytchison, who was just twenty-one at this time, was the man's employer at Pacific Palisades Furniture Company. Even though he was arrested for suspicion of theft, said Sperry upon his safe landing and arrest, "I didn't steal the plane. I had flown it several times. I taught Bill to fly it."[116] Eytchison defended Sperry, calling him a good salesman who was simply a victim of his wartime experience.

Fast forward many years later, to 1990, when Sperry coauthored the book *China Through the Eyes of a Tiger*, about his experience as a Flying Tiger in WWII. (Coincidentally, the 1963 account from Mr. Wykoff claiming to have met Amelia Earhart also came from a self-proclaimed Flying Tiger.) As it turns out, the book was a mostly fabricated account of Sperry's experience during WWII. Sperry even used a picture of a deceased fellow pilot in his book, claiming to be him. According to a May 23, 1999, article from the *Orange County Register*:

> This week, after months of inquiry by *The Orange County Register*, Sperry admitted that he's been living a lie. One he profited from. One that

---

[116] "Unhappy Pilot Lands Safely After a 4-Hour Suicide Hop," Associated Press, *Battle Creek Enquirer*, March 28, 1953, p. 1.

```
veterans say cheapened their service. He never
piloted a military plane. There were no downed
fighters. He wasn't an officer. His book — "I was a
fighter pilot, a Flying Tiger. There were guys who'd
give their eye teeth for my job, with good reason"
— is all fiction. "I don't know why I did it," said
Sperry, who turned 77 on Friday. "I'm ashamed. I
should have stopped it, so help me God."[117]
```

In short, this was the man who taught Bill Eytchison to fly. Eytchison's own statements appearing in news articles following the crash of 51-Poppa indicate that the purpose of the flight from Vegas to Long Beach with Joan had been to discuss a future, record-setting pole-to-pole flight that would take place in September 1965. The plan was that Joan would fly the plane, and he would be her copilot. However, as noted at the outset of Chapter 7, Joan had already written him off at this time, believing this polar flight would never become a reality. One oddity stands out: two days after Joan's fatal plane crash, Eytchison is quoted as saying that he had "bought a $117,000 plane just two days ago."[118] Would it be a coincidence that he happened to buy this plane on the *very day* that Joan died?

If the crash was not an accident, it seems logical to ask, who would want the plane to go down? Did Eytchison have enemies? Would someone benefit from cutting Joan, or both of them, out of the picture? Is it possible that, in buying the plane on the day Joan died, Eytchison's plan had been to embark on a polar flight all along, with or without Joan, and just to use Joan's name in order to attract interest? Was money owed to a disgruntled sponsor, or to someone in Vegas? (This was a time in history when what happened in Vegas really did stay in Vegas.) Could it be that a competitor had caught wind of plans for the pole-to-pole flight

---

[117] "Roland Sperry 'Fesses Up," *Orange County Register*, May 23, 1999, https://www.warbirdforum.com/sperry.htm.

[118] Harry Trimborn, "Experts Seek Cause of Women Fliers' Crash," *Los Angeles Times*, February 19, 1965, Main Edition, p. 33.

and wanted to play hardball? Or was this simply all bad luck? Was it just that Joan was flying an old, problematic plane that had already run into many mechanical issues?

Did the Competition Ever Get Dirty?

The 1960s was a vibrant period full of aviation firsts. For example, in March 1961, Max Conrad completed an around-the-world flight at the equator—from west to east—in a twin engine Piper Aztec in just eight days, eighteen hours. He was accompanied by Richard Jennings, an observer for the record flight. In October 1961, Jacqueline Cochran set two Fédération Aéronautique Internationale (FAI) world records, taking a Northrop T-38A-30-NO Talon supersonic trainer to an altitude of 55,252 feet in horizontal flight and reaching a peak altitude of 56,072 feet.[119] In May 1963, Betty Miller became the first female pilot to fly solo across the Pacific Ocean.[120] Then, in 1964, there were Joan Merriam Smith and Jerrie Mock's world flights, with each setting a number of new records in her own right. In May 1964, Cochran next set a speed record of 1,429 miles an hour in a Lockheed F104G Super Starfighter at Edwards Air Force Base.[121]

In mid-November 1965 (about two months after Eytchison had initially planned/hoped to complete his and Joan's flight, and nine months after Joan died), a pole-to-pole flight carrying thirty American scientists and medical and aviation experts was completed in the name of scientific exploration. It was sponsored by Col. William F. Rockwell of Rockwell-Standard, and two pilots from TransWorld Airlines—Captain Fred Austin and Captain Harrison Finch—headed the crew. Of note, Lowell Thomas Jr. (whose father was to write the foreword for Trixie's book) was also aboard, along with Arctic pilot and explorer Bernt

---

[119] "12 October 1961," *This Day in Aviation* blog, October 12, 2018, https://www.thisdayinaviation.com/tag/t-38a-30-no/.

[120] "Betty Conquers Pacific Skies," Associated Press, *The Post-Standard* (Syracuse, NY), May 13, 1963, p. 3.

[121] Richard West, "Air Speed Record Set by Jacqueline Cochran: World Mark for Women of 842.6 m.p.h. Achieved in Supersonic Jet Trainer," *Los Angeles Times*, August 26, 1961, Main Edition, p. 12.

Balchen and Robert Prescott, president of the aptly named Flying Tiger line (the first scheduled cargo airline in the United States). Prescott's company was the owner of the Boeing 707-323 plane that completed the route. This became the first flight in history to circumnavigate the globe via the North and South Poles.[122]

In short, these circles ran tight, and there was certainly a level of friendly competition when it came to conquering these various aviation feats. But the question is: did the competition ever get dirty? After Jerrie took off for her world flight, for example, it is famously known that she learned that her radio did not work during her first over-water crossing of the Atlantic. In her book, when describing the situation upon landing in Bermuda, she recalls:

> In less than a minute he sat up and looked at me with a funny expression on his face. "Well I found the problem, and it'll be easy to fix. But I sure don't understand the men who installed the radio," said the radio man in Bermuda. He continued, "Well, there's a wire disconnected, all right. One to the control head. And it didn't just come off—the raw lead is all taped up and tucked away. The radio could never work like that."[123]

Wondering how that could have possibly happened, Jerrie speculated that the disconnection must have occurred in Wichita—not Florida, where it had been installed—because the radio had just been checked and no one could have gotten to it afterward. However, it wouldn't be until many years later that she would feel comfortable attributing this incident to Joan. In her aforementioned Buzzfeed article, Amy Saunders writes, "After 50 years, Mock figures she can say now what she couldn't then: sabotage. 'It definitely was not accidental,' she says. Someone, she

---

[122] "Polar Round-World Plane Ends," *Victoria Advocate* (Victoria, Texas), November 18, 1965, p. 6.

[123] Jerrie Mock, *Three-Eight Charlie: 1st Woman to Fly Solo Around the World*, Fourth Edition, Kindle Edition (Granville, Ohio: Phoenix Graphix Publishing Services, April 10, 2014), Chapter 3.

thinks, wanted Smith to win the race." Initially, upon reading this, I thought that someone could have just as easily messed with the wiring in 51-Poppa, even if, or especially if, it had just been checked.

What Really Caused the Cessna 182 Crash?

On February 17, 1965, Joan and Trixie took off from Long Beach in a Cessna 182—destination unknown. Shortly thereafter, their plane crashed into the San Gabriel Mountains near Wrightwood, California. Following conversations with Shirley Thom—a fellow Ninety-Nines member who had known Joan and Trixie personally in 1965—at first I find nothing suspicious regarding the crash of the Cessna 182, other than the fact that everyone was extremely shocked. According to most newspaper accounts (as well as the account of Joan's husband, Jack Smith), Joan was up in the air that day to test fly a plane equipped with turbochargers for the Rajay Corporation, by whom she was employed as a test pilot. This gives us a good place to start.

In a January 27, 1965, journal entry (about halfway between the crash date of 51-Poppa and the Cessna crash), Trixie wrote about spending time up in the air with Joan to test turbochargers. "Wednesday a.m. Joan called and asked if I'd like to fly with her. We were up three hours in a Cessna 182 experimental plane, at 15,000 feet putting time on the test plane equipped with turbochargers. She used computer and wrote data for turbos at various power settings, etc. WONDERFUL DAY."[124] If you do the math, January 27 was only about two and a half weeks after Joan's crash landing in the desert.

As fate would have it, Trixie actually was not supposed to be up in the air on the morning of February 17. According to a *Long Beach Independent* article dated February 18, 1965, Joan's close friend Joyce was supposed to have joined her on this particular flight, but she had declined because of her home's dustiness.

> Joyce Dixon, Long Beach publicist and close friend of Joan Merriam Smith is alive today because her

---

[124] Trixie Schubert's personal journal, January 17, 1965.

```
home was "a bit dusty" on Wednesday. At lunch
Tuesday, Mrs. Dixon said, "Joan invited me to go
flying with her the next day. I told her I couldn't
go, that my house was a bit dusty. 'Why don't you
take Trixie along,' I asked her."[125]
```

Trixie had been up in the air before with Joan to test turbochargers; they both loved flying, and they were working on a book together. But where was their actual destination that morning, whose plane were they flying (i.e. did Rajay own the plane, or was it leased through someone else?), and what really caused it to fail? In newspaper articles, Bill Eytchison is quoted as saying that he believed the pair was en route to Lucerne Valley to view the crash site of 51-Poppa, in order to gather information for Joan's book. Shirley Thom had always remembered hearing that they were on their way to Vegas. No flight plan was filed.

The Cessna 182 plane crash was well publicized: several hundred news articles ran regarding the crash in the month following the accident. Newspaper accounts largely suggest that the crash was caused by everything from structural failure to pilot error. However, the vast majority of newspaper coverage in the month following the crash refers to the official National Transportation Safety Bureau (NTSB) crash report, which listed the probable cause as "pilot exceeded designed stress limits of aircraft" and/or "miscellaneous acts, conditions—overload failure." But because there was disagreement with the preliminary analysis, a second investigation was forced.

I made personal efforts to obtain the full narratives on each crash through various organizations. After repeated check-ins with the NTSB, the email response I received was simply, "There are no narratives for these cases. Accident dockets have been destroyed." So let's turn to the evidence we do have.

To begin with, you may recall from the beginning of this chapter how Jack Smith mentioned that the cause of Joan and Trixie's crash had

---

[125] "With Dusting to Do, Friend Didn't go on Fatal Air Trip," *Long Beach Independent*, Vol. 27, No. 147, February 18, 1965, p. 7.

to do with a folded wing. According to the recollections of Trixie's children, there was a court case immediately following the Cessna crash whereby the surviving families sued the Rajay Corporation, which was, of course, Joan's employer at the time. The memory is that a settlement was reached based on the finding that the plane had been involved in a prior accident, after which its right wing had been previously repaired. Upon hearing this, I immediately wondered whether this had just been a convenient settlement to quickly shut down a high-profile case? However, because I was unable to locate a historical court case to this effect, I cannot say with certainty that this is what actually transpired.

But I was able to find a copy of a deposition for a different court case whereby Trixie's husband later sued the life insurance company in 1968.[126] In that deposition, a reference was made to another case when Delwyn Schubert was asked about whether or not he "gave any statements to the insurance adjuster from the companies involved in the other action." In this deposition were many questions about Trixie's unpublished book, the files she kept in her office, the notebooks and tape recordings she took, the book's contractual status, the copies of the book she kept, the history of her friendship with Joan, what the children knew about Trixie and Joan's friendship, details about how the particular flight materialized on February 17, the nature and findings of Joan and Trixie's husbands after visiting the crash site in Wrightwood, the understanding of Joan's relationship with Rajay (referred to Rajaw in the deposition), and more. Ultimately, for me at least, these post-crash conversations only raised more questions.

A Single Witness Account

Let's now turn to examining some of the evidence collected and observations made as to the real cause of Joan and Trixie's Cessna crash. A single witness account (see Appendix E) may help to provide some clues. On February 21, 1965, at three-thirty p.m., Mr. Jones, resident

---

[126] "Delwyn George Schubert (Plantiff) vs. The Mutual Life Insurance Company of New York, et al. (Defendants)," Case # 878-835, Superior Court of the State of California for the County of Los Angeles, November 21, 1968.

manager of the Southern California Bible Conference,[127] gave the following account to Mr. William Moore, air safety investigator with the Civil Aeronautics Board Bureau of Safety:

> I was out on the conference grounds and happened to look up at the sky and observed a jet stream and was following it with my eye when I noticed a small plane. Then I returned to following the jet stream with my eyes. Then I heard the motor of the small plane rev up, so I turned my head to look at it. I noticed that the right wing had folded back against the plane. Then the plane went into a nosedive and crashed.
>
> It was a clear day. The small plane was flying high above the mountains. I would estimate about 2,000 feet. When I first observed the small plane it was in no trouble at all, flying in a normal manner headed southwest. When I continued to follow the jet stream, my attention was turned back to the small plane because of the motor being revved up.
>
> I called my wife to come and spot the smoke from the crashed plane with me, that we would be certain of the location. Then we called the fire department and the Forestry Service at Big Pine.
>
> I got in my car and drove immediately to the scene of the crash. There was one man there when I arrived. I do not know his name. I do not believe that he had seen the plane in the sky, only the smoke and fire caused by the crash. Soon the Fire Department arrived and I helped them put out the fire.[128]

---

[127] The Southern California Bible Conference is still in operation today and has since been renamed the Verdugo Pines Bible Camp.

[128] "Account of Witness to Airplane Crash of Trixie-Ann Schubert and Joan Smith," an account taken by Mr. William G. Moore, Air Safety Investigator, Civil Aeronautics Board of Mr. Robert L. Jones of Wrightwood, CA, February 21, 1965.

Mr. Jones also relayed to Mr. Moore that the plane was not on fire in the sky and estimated it was about ten to twelve seconds from the time he heard the motor of the plane rev up and saw the right wing folded back against the plane until it nose-dived and crashed. In the years following the crash, Trixie's children recalled having heard that Mr. Jones had been trained as a plane spotter for the military. During World War II and the subsequent Cold War, some countries encouraged their citizens to become "plane spotters" as part of an observation corps for reasons of public security. The level of detail given in his personal account helps lend even more credibility to the accuracy of his recollections.

An Online Forum about Wrightwood History

Building upon this witness account, in 2007, an individual by the name of Mr. V. Koenig shared his ideas about the crash via a Wrightwood, California, online forum in response to an article that was written about the historic crash by Terry Graham, a Wrightwood local. Reflecting on the fact that he had met Joan in 1964 while stationed at the U.S. Naval Air Station in Agana, Guam, Mr. Koenig wrote:

> For years, I've tried to find out more information regarding the circumstances of her death and her subsequent burial location. I researched the NTSB records and found two. The first had to do with the crash of her record-setting Piper Apache which was lost near Lucerne Valley, California, and the other having to do with the incident which resulted in her death. The NTSB reports are a cold and brief description of an aircraft incident/accident, injuries/fatalities and a sterile summary of probable cause. In this instance, Joan Merriam Smith was faulted for "pilot error."
>
> I don't believe the NTSB analysis and summation of this accident for one minute. Joan Merriam Smith was an excellent pilot. In addition to her private pilot, commercial pilot, instrument pilot and multi-engine pilot ratings, she also possessed her airline transport pilot rating. She knew what she was doing while at the controls of ANY airplane. I firmly believe that Joan Merriam Smith

and her unknown to me biographer encountered sudden extreme clear air turbulence which exceeded the structural capability of the aircraft in which she was piloting... I believe she encountered sudden extreme clear air turbulence and perhaps did not have sufficient time to reduce power and ride out the bumps.[129]

When taken together, these two accounts generally line up with the premise that turbulence likely caused a weak wing to fail. But what, besides clear-air turbulence created while flying over the mountains, might cause a plane to break apart on an otherwise clear and sunny day? What was the precise forensic progression of failure? What exactly would cause a motor to rev up, and was it the motor that Mr. Jones actually heard? And where did that jet stream come from?

A Deeper Investigation

In May 1965—three months following the crash—*CAP (Civil Air Patrol) Times* published an article entitled "Search for Aviatrix's Plane Ends on California Peak." The article provides some detail on the search for missing plane parts following the crash, but it doesn't fully explain what was found, or where.[130]

Multiple entities were involved in the post-crash investigations, from the local forest service and the Civil Aeronautics Board (CAB) to the Civil Air Patrol and the FAA. To collect data, investigators used pack animals, search teams, vehicles, and telescopes to locate evidence over the course of a month (due to weather delays). The article says that Mr. William Moore (who took the single witness account) showed great interest in the findings, noting that investigators had never been able to locate the missing pieces of the tail and wing from the Cessna. In short, the article

---

[129] Wrightwood, CA Public Forum, "A Piece of History, Legend Down," July 3, 2007, http://www.wrightwoodcalif.com/forum/index.php?topic=9329.0.

[130] The beginning of my copy of this article has been cut off. In searching for the full text, I was able to locate 1960s archives online at the *CAP Times* history page (https://history.cap.gov/document/49), but there are no issues from 1964 or 1965 available.

regarding the post-crash investigation focuses primarily on how the search was conducted but doesn't explain exactly what was found. It suggests that three months later the authorities were still open to interpretations about what actually happened based on the pending collection of more evidence.[131] Locating the missing plane pieces would be critical to forming a scientific explanation for what may have occurred versus having to make an educated guess.

Eight months following the Civil Air Patrol's article, however, an Associated Press article helped to clarify the findings of the second investigation. Specifically, it referenced the conclusion made by the Civil Aeronautics Board.

> Joan Ann Merriam Smith, first woman to make a round-the-world equatorial flight alone, was killed in a plane crash last February because she encountered turbulent air while flying at high speed, the Civil Aeronautics Board said today. The resultant violent upward and downward forces broke off the tips of both wings and part of the tail structure of the Cessna 182 that Mrs. Smith was flying and sent it hurtling into a 7,260 foot high ridge near Wrightwood. Calif., Feb. 17.[132]

While I was unable to locate the original CAB report, a March 1966 issue of *The AOPA Pilot* also provides some additional detail about the CAB's report of findings in the sidebar of an article entitled "The Fiery End of 51 Pops." Here's an excerpt:

> The summary CAB report of the fatal accident, issued recently, said that the flight was made "in

---

[131] "Search for Aviatrix's Plane Ends on a California Peak," *CAP Times*, May 1965, p. 1 & 16.

[132] "Turbulent Air at High Speed Cause of Crash," Associated Press, *Reno Gazette Journal*, February 10, 1966, p. 11.

> conjunction with functional and reliability tests of the experimental turbocharged engine installed in N8784T." It added: "The flight was to operate at or above 5,000 feet m.s.l., with turbocharger engaged and the pilot was to investigate climb performance not to exceed 23,000 feet, m.s.l. Clear weather conditions prevailed throughout the proposed flight area and no flight plan was filed . . . Witnesses differ as to the direction of flight at initial sightings but agreed the aircraft was flying at 1,000 to 1,500 feet above the terrain. Shortly thereafter they saw N8784T with the wings failed crash out of control and burn at the 7,260-foot elevation of a 50 degree upward slope on Blue Ridge while on a northerly heading [. . .]
>
> There was no evidence of fatigue or evidence of malfunction or failure of the aircraft prior to the in-flight structural failures.

For what it's worth, this summary differs slightly from the single known witness account, which refers to the fact that both wings failed (not one), that the plane was headed to the north (versus the southwest), and that it was flying between 1,000 and 1,500 feet above terrain (not 2,000 feet). It also establishes that no flight plan was filed and that there was no evidence of malfunction prior to structural failure. These contradictions are significant because if only one wing failed, this would support the idea of a faulty wing, but if both wings (and the tail) failed, this could indicate something more serious.

Of course, it is possible that the Cessna hit a patch of clear-air turbulence above the San Gabriel Mountains. Clear-air turbulence (CAT) is defined as the turbulent movement of air masses in the absence of any visual clues such as clouds; it is caused when bodies of air moving at widely different speeds meet. According to atmospheric physicist and forensic meteorologist Dr. Elizabeth Austin, in her book *Treading on Thin Air: Atmospheric Physics, Forensic Meteorology, and Climate Change: How the Weather Shapes Our Everyday Lives*, weather accidents comprise more than 21 percent of aviation accidents each year, and the number one culprit in weather-related accidents is wind. In her book, she writes, "Wind is a factor in 48 percent of these cases. It can cause crashes due to very strong winds or wind shear (changes in wind speed and/or direction over a very short distance)."[133]

Upon hearing this, I wanted to learn more about this phenomenon. After watching a 2018 episode of the TV show *Aircrash Confidential*, in which Dr. Austin was interviewed, I learned more about the unique wind conditions created by the Sierra Nevada mountain range. Because of the way the range is shaped and the fact that it runs almost perpendicular to the jet stream, wild and sometimes unpredictable wind conditions are often created, even on clear and sunny days. In the episode, Dr. Austin talks about the concept of the "Nevada Triangle," an area between Reno, Las Vegas, and Fresno, CA, that is known for having an unusually high number of plane crashes. While searching for the adventurer Steve Fossett's missing plane in the Sierra Nevada during 2007, for example, investigators discovered remnants of some two thousand small plane crashes over a sixty-year period.

In a National Public Radio article about the Nevada Triangle, Desert Research Institute climatologist Kelly Redmond explains, "So, when the winds are coming up, they come up kind of smoothly on the west side and then they have this very rapid descent and there is a tremendous amount of turbulence that can be caused by such situations."[134] Redmond explains that for small planes, which must fly lower than commercial jets and at less power, getting caught in that air flow can make it very difficult to get out. It's comparable to getting caught in Niagara Falls.

---

[133] Elizabeth Austin, Ph.D. *Treading on Thin Air: Atmospheric Physics, Forensic Meteorology, and Climate Change: How the Weather Shapes Our Everyday Lives*, (New York, NY: Pegasus Books, Ltd, April 2016), p. 190.

[134] Joe Schoenmann, "The Nevada Triangle: A Graveyard For Planes," *National Public Radio*, September 9, 2015, https://knpr.org/knpr/2015-09/nevada-triangle-graveyard-planes.

A famous example of this specific phenomenon occurred in 1976 when a Cessna 182P crashed near Mt. Whitney—the highest summit in the contiguous United States—and the lone survivor lived to tell about it. In the same *Aircrash Confidential* episode, aircraft accident historian G. Pat Macha explains how the Cessna 182P was compromised by an unexpected downdraft, and how the plane should ideally have been flying at a little higher altitude above the ridgetops. Having visited more than eight hundred crash sites, Macha has not only written the five hundred-plus-page book *Aircraft Wrecks in the Mountains and Deserts of California 1909-2002*, but he also has given countless lectures across the country since first beginning his career in 1963.

In terms of the Cessna 182P crash that was ultimately caused by a downdraft, two of the Cessna's three passengers died shortly after impact. At this point, Lauren Elder—who had been up in the air as a photographer that day to take photos of the Sierra—decided it was time to take her fate into her own hands and embark on the treacherous descent off the mountain. Describing Elder's journey to safety in the TV show, Macha explains the difficulty and perseverance it took for Elder to drag herself off the mountain in order to survive: "There is no trail. It is brutal terrain. Vertical. Steep. As she descended, she was hallucinating at points. Were they people? Did she see a house? But she kept going. She kept that relentless descent, lower and lower, warmer and warmer."[135] Sustaining life-threatening injuries, Elder ultimately scaled down some 9,300 feet in elevation, and after thirty-six hours of no real sleep or food, she stumbled into the mountain town of Independence, California. After her arrival at a local motel, her story at first was not taken seriously. In her book, she describes the reaction of the motel's front desk attendant upon noticing her disoriented condition: "He was sitting in the office of a motel, watching television, but he was not going to help me. I told myself I should not

---

[135] "The Nevada Triangle," Aircrash Confidential, a television series produced by WMR Productions and IMG Entertainment, produced by Matthew Barrett, David McNab, Alan Griffiths, Season 3, Episode 5, original air date August 3, 2018.

be surprised."[136] Eventually, she was picked up by local authorities who had heard of an earlier crash on the radio and was subsequently taken to the hospital. Because so few people have lived to tell about the experience of unexpected downdraft, Elder's story offers insight into the risks and reality of the unpredictable wind conditions over California's mountain ranges.

While Joan and Trixie's crash occurred in the San Gabriel Mountains, a mountain range located just south of the Sierra Nevada, due to the similar physical terrain, it is entirely possible that this same weather phenomenon ultimately affected them too. After all, it is believed that Steve Fossett—who also was an experienced world solo pilot—died in a fatal plane crash over the Sierra Nevada due to a forceful and unexpected downdraft. Coincidentally, he also was flying a Cessna, he hadn't filed a flight plan on the day of his fateful crash, and the weather conditions that day were sunny and clear. But is it also still possible that something else could have caused their crash?

Possible Jet Interference

Since WWII, planes from Edwards Air Force Base, El Toro Marine Corps Air Station, and Lockheed's famous Skunk Works flight test facility at Palmdale would routinely run test flights and practice combat maneuvers near the San Gabriel Mountains. In 1943, for example, a military test plane crashed right in the middle of the town of Wrightwood.[137] In May 1964, Jacqueline Cochran set a speed record just outside of Edwards Air Force Base, hitting 1,429 miles an hour in a Lockheed F104G Super Starfighter.[138] And in May 1965, the Lockheed YF-12A—a precursor to the SR-71 Blackbird—shattered nine records

---

[136] Lauren Elder and Shirley Streshinsky, *And I Alone Survived*, (New York, NY: E P Dutton; 1st edition, March 1, 1978), p. 171.

[137] Terry Graham, "A Piece of History, Legend Down," Big Pines Volunteers Newsletter of Wrightwood, CA, September 2011, Vol 3, Issue 9. http://www.wrightwoodcalif.com/bigpines/Newsletter%20September%202011.pdf.

[138] Dorothy Cochrane and P Ramirez, "Jacqueline Cochran," Smithsonian National Air and Space Museum, accessed August 8, 2018, https://airandspace.si.edu/explore-and-learn/topics/women-in-aviation/cochran.cfm.

in one day of testing at Edwards.[139] (The SR-71 set many records such as hitting a sustained speed of 2,070 miles per hour and an altitude of 80,257 feet.)

In theory, being either preceded (in its flight path) or bypassed unexpectedly by a plane could create enough turbulence to cause a small Cessna to break apart and crash. I first heard of this as a possible cause through my interactions on the Internet. After exploring this idea with my network, I learned that this phenomenon is not entirely uncommon. Similarly to boats on a lake creating wakes that can affect other boats in their proximity, passing planes also can leave behind turbulent wakes that have the capacity to affect nearby planes. Macha says a notable example of this type of crash occurred in 1993 when a private jet carrying two top executives of the In-N-Out Burger chain inadvertently got caught in the wake of a Boeing 757 and crashed.[140]

While these types of incidents more commonly occur closer to the ground, according to an article from the Institution of Mechanical Engineers about the latest research on the wingtip vortex (the primary component of wake turbulence), wingtip vortices pose a serious danger to smaller aircraft, generating tiny horizontal tornadoes that trail behind large airplanes. Specifically, "wingtip vortices generated by large, bypassing aircraft can create enough turbulence to flip a plane over."[141] Wake turbulence caused by larger, heavier aircraft also can descend or "sink" as they are created, meaning that turbulence created by a larger aircraft can also descend from the airspace above to affect a smaller plane flying below.

---

[139] "NASA Armstrong Fact Sheet: YF-12 Experimental Fighter-Interceptor," NASA website, February 28, 2014, https://www.nasa.gov/centers/armstrong/news/FactSheets/FS-047-DFRC.html.

[140] Rebecca Trounson and Jeffrey A. Perlman, "Crash That Killed In-N-Out Officers Is Investigated," *Los Angeles Times*, December 17, 1993 http://articles.latimes.com/1993-12-17/news/mn-2813_1_fiery-crash.

[141] Joseph Flaig, "'It can flip you over': Wing shape research could reduce dangerous vortex turbulence," Institution of Mechanical Engineers, May 15, 2018, http://www.imeche.org/news/news-article/%27it-can-flip-you-over%27-wing-shape-research-could-reduce-dangerous-vortex-turbulence.

According to my email correspondence with a former USAF pilot who knew both Joan and Trixie, a near or actual mid-air "hit and run" is a valid possibility. Drawing upon his own experience as a pilot, he was kind enough to walk me through several scenarios:

The last possibility that comes to mind is a near or actual mid-air "hit and run" by a test aircraft in which the military or accident-causing aircraft might not have been seriously damaged, if at all, and was therefore easily capable of continuing its flight, and of safely continuing on to its home base—say at any of the nearby airfields. Additionally, if this occurred, the military jet pilot might not have even been aware of what he might have caused, particularly had he either flown near the small Cessna or ahead of its path. Or, more likely, if this actually happened, the high speed jet would not have had to actually impact the Cessna, but either have passed too close to it, or have previously generated a wingtip vortex of sufficient magnitude to have caused structural damage to the Cessna, especially had the turbo-charged small airplane been cruising at high speed.

If any of these dangerous possibilities actually occurred, then there might be an issue of latent liability for this accident—by other than Joan—or Rajay.

It may be worthy to note that test pilots, in particular, tend to have their "heads inside the cockpit" much of the time, in order to check the numerous extra flight test instrumentation they need to observe during their test flights, thereby not watching outside, as a pilot would normally do, in order to see any other air traffic. This is one of the reasons why most test flights are done in "RESTRICTED" areas, which actually prohibit the presence of other civilian or military air traffic.

At this point, you may recall from the single witness account that Mr. Jones referred to spotting a "jet stream" that he followed across the sky before noticing the small Cessna. The technical term he was most likely referring to is an exhaust contrail. According to the NASA Langley Research Center, a contrail is formed by water vapor condensing and freezing on particles from airplane exhaust. Contrails are always made of ice particles, due to the very cold temperatures at high altitude.[142] Contrails—not to be confused with chemtrails—are otherwise known as the white lines we see across the sky created by passing planes. The question is: was the contrail Jones spotted created by the Cessna, or was it a passing plane? Because a plane needs to be flying at a high altitude (usually about 25,000 feet) in order to produce a visible contrail, it's unlikely to have been created by the Cessna. But if it was a passing plane, could it have passed by close enough to affect the stability of the Cessna? Or, even if it wasn't the plane that created the contrail, could another plane (commercial or military) possibly have flown past just before and created enough turbulence to destabilize the Cessna?

According to a 2016 report about wake turbulence published by the Civil Aviation Authority of New Zealand:

> The most dangerous situation is when a small aircraft flies directly into the wake of a larger aircraft. This usually occurs while flying beneath the flight path of the larger aircraft. Flight tests conducted in this situation have shown that it is not uncommon for severe rolling motions to occur with loss of control. In other instances, if the aircraft is flown between the vortices, high roll rates can coincide with very high sink rates in excess of 1,000 feet per minute. Depending on the altitude, the outcome could be tragic.[143]

---

[142] "Contrail Science," The Contrail Education Project, NASA Langley Research Center, last updated 3/20/18, https://science-edu.larc.nasa.gov/contrail-edu/science.html.

[143] "Wake Turbulence," Civil Aviation Authority of New Zealand, last modified April 2016, p. 4, https://www.caa.govt.nz/assets/legacy/safety_info/GAPs/Wake_Turbulence.pdf.

In other words, if Joan and Trixie unexpectedly crossed the wake created by a larger, slower-moving aircraft passing above them, then it's entirely possible that their plane could have been immediately destabilized by such an incident. According to the FAA, wake turbulence was not generally regarded as a serious flight hazard until the late 1960s, when the phenomenon was first discovered.[144]

Regardless of what actually occurred—whether it be faulty plane parts, poor maintenance, clear-air turbulence, a near-air collision with another plane, actual pilot error, or a combination of these or other things—the fact remains that no one really knows for sure what actually happened. We can only speculate.

---

[144] Bruce Lansford, "Wake Turbulence: Should You Worry? This Invisible Hazard Continues to Threaten Light Aircraft," October 5, 1998, Aircraft Owners and Pilots Association, https://www.aopa.org/news-and-media/all-news/1998/october/pilot/wake-turbulence-should-you-worry.

= CHAPTER 9 =

JUSTICE FOR JOAN

"No difficulty can discourage, no obstacle dismay, no trouble dishearten the man who has acquired the art of being alive. Difficulties are but dares of fate, obstacles but hurdles to try his skill, troubles but bitter tonics to give him strength; and he rises higher and looms greater after each encounter with adversity."

— Ella Wheeler Wilcox

Upon completion of her world flight, Joan became the first person in history to fly solo around the world at the equator; the first person to complete the longest single solo flight around the world; the first woman to fly a twin-engine aircraft around the world; the first woman to fly the Pacific Ocean from west to east in a twin-engine plane; the first woman to receive an airline transport rating at the age of 23; the youngest woman to complete a solo flight around the world; and the first woman to fly solo from Africa to Australia, from Australia to Guam via New Guinea, and from Wake to Midway Island.[145] After 27,750 miles and 56.5 days, she had finally completed her goal of finishing what Amelia Earhart had started. None of Amelia Earhart's originally scheduled stops were missed, with the exception

---

[145] "Joint Resolution in Honor of Joan Merriam Smith," *Ninety-Nine News,* June 1965, p. 6, https://www.ninety-nines.org/pdf/newsmagazine/196506.pdf.

of Rangoon, which she overflew since landing there was forbidden unless in the case of an emergency. At the end of her flight, Joan had collected more than eight hundred signatures on her plane from various people around the world.

According to information about Joan's flight from the Smithsonian Air and Space Museum, 10,550 miles were crossed over sea and 17,250 miles over land.[146] For approximately 45 percent of the trip, there was no air traffic, with little to no radio aids available. In many instances, Joan had to forecast her own weather. She flew within one thousand miles of the equator during her trip, crossing the equator four times, and dealt with two equatorial fronts.[147] Between Darwin, Australia, and Lae, New Guinea, for example, Joan battled the most unusual, violent weather in her flying experience, later discovering that she had even flown through the partial formation of a cyclone. Facing headwinds ninety percent of the time, this resulted in an average groundspeed of 120 knots (138 miles per hour) for her while doing 155 knots (178 mph).[148] All told, there were some thirty days lost to unforeseen weather delays, mechanical problems, or unexpected incidents.

By comparison, Jerrie Mock covered 23,103 miles in 29 days, 11 hours, 59 minutes (but who's counting)?[149] Jerrie's flight took her across the Northern Hemisphere. She not only accomplished her goal of becoming the first woman to fly around the globe solo, but she also set a record for speed. On May 4, 1964, President Lyndon B. Johnson publicly awarded her the FAA's Gold Medal for Exceptional Service at the White House. Notably, she was also the first woman to receive the Louis Bleriot Silver Medal from the Fédération Aéronautique Internationale.

---

[146] McNabb.

[147] John Sarver, "Joan Merriam/Around the World Flight Facts," a Sarver & Mitzerman press release, May 1, 1964, courtesy 99s Museum of Women Pilots.

[148] Joan Merriam, "Tribute to a Star: Flying the A.E. Route."

[149] "Geraldine Mock Collection 1963-1964," Smithsonian National Air and Space Museum, accessed September 23, 2018, https://airandspace.si.edu/collection-objects/geraldine-mock-collection-1963-1964.

Because Joan felt that she was never given a fair chance to set the record, combined with the fact that her historically significant route was also much longer and more difficult than Jerrie's, she felt slighted by a lack of acknowledgement, as did her supporters. When pressed, FAA Administrator Jeeb Halaby expressed that he felt it was a moot point. The following excerpt, drawn from a communication between Joan and Halaby, appears in Gilbert's "The Loser":

> "'I am sorry that you feel [the world flight] has not received sufficient recognition by the Federal Government. We have received many letters from political and public relations sources' (and you can almost hear him adding under his breath, 'and we didn't need any of them') 'and we have reviewed all the circumstances surrounding it. However, it is my opinion, as Administrator of the Federal Aviation Agency that the feat speaks for itself, has been recognized by me and my associates, and will be judged by history. This is more important than further action from the Federal Government.'"[150]

Because Jerrie had received the majority of the fanfare, it was clear that Joan felt a bit cast aside. It also was clear that the FAA was not interested in doing anything more about it. From their point of view, this was a black-and-white issue: Joan did not get a sanction, and she did not finish first. Perhaps this was a headache that they just weren't willing to deal with. But from the view of many others outside the FAA, it was all shades of gray.

A Solid Crew of Support

As noted by Mr. Halaby, many letters of support for Joan from both political and public relations sources would come pouring in. Among the letters sent to various organizations from multiple individuals on

---

[150] Gilbert, p. 81.

Joan's behalf was an April 20, 1964, letter from John Sarver to Ruth Deerman, international president of the Ninety-Nines. In an excerpt from the four-page, single-spaced, typed letter, Sarver explains the basis of the perceived injustice. He writes:

> *Admittedly, I am partial to Joan Merriam—but I am also active in sports and understand rules, regulations, and fair play. In the case of Joan Merriam she received a very unfair decision. To illustrate this point: if a race were being run around a track and one runner shortcut across the middle and claimed first place, would he be considered the winner? Of course not! The first requisite is that each runner cover the full distance. If a person were to fly around the world officially, wouldn't the first requisite be that he or she fly the distance, 24,950 statute miles?*

Later in the letter, Sarver adds,

> *The NAA sanctioned Mrs. Mock for "speed around the world feminine." The fact that they sanctioned a route shorter than the required miles to circle the globe is one question. The fact that they "conveniently" allowed her to miss three required stops then "make-up" the mileage by "flying around" in the United States is another question....*

> *The world speed mark set by Jerrie Mock should be contested if one is ever again to believe aircraft records. Mrs. Mock did set the first northern circumference of the world speed record for a woman. This is true! As for this record being "official" is questionable. Ground rules were broken! In baseball you cannot hit a ball over the fence and make a home run without touching all bases! In golf you cannot win a tournament without sinking every putt, regardless how near the hole your ball is! Every sport has its rules and by the rules the game must be played.*[151]

---

[151] Letter from John Sarver to Ruth Deerman, April 20, 1964.

On September 3, 1964, aviation heavyweight Jacqueline Cochran would next lend her support. An accomplished aviation pioneer, Cochran was not only appointed the director of the Women Airforce Service Pilots (WASP) in 1942—the U.S. Army Air Forces program that tasked some 1,100 civilian women with noncombat military flight duties during World War II—but she set many dozens of aviation records in her own right and was the first woman to break the sound barrier. In a letter to Randleman at the NAA, she stated, "So far as I know you were fair and followed the rule book . . . But," she continued, "I want to see that she gets all she is entitled to without taking anything away from Mrs. Mock." Specifically, Cochran asked two questions: 1) Had the NAA ever certified or made official a record that was not previously sanctioned, and 2) Was there a category under feminine records for Class CID (3,858 to 6,614 pounds) for around-the-world flights?[152] In other words, she was trying to determine if the NAA would consider finding some other way—still fair—to recognize Joan's accomplishment more formally.

Six days later, the NAA's executive director would issue a return response. On September 9, 1964, Colonel Mitchell Giblo explained that the NAA could not certify a record that had not been previously sanctioned.[153] He reminded her that, due to regulations, the organization could only sanction one contestant at a time for a speed record.[154] (Jerrie had obtained a sanction for a feminine speed around-the-world Class CIC, or 2,204-3,858 pound record. But, technically, there was no feminine speed record available for Class CID, or 3,858-6,614 pounds.)

On November 1, 1964, Trixie also wrote to Ruth Deerman. In her letter, she outlined the various reasons why she felt Joan deserved

---

[152] Letter from Jacqueline Cochran to M.J. Randleman, September 3, 1964, National Aeronautic Association Archives, Box 139, Folder 6 (Acc. XXXX-0209). Archives Department, National Air and Space Museum, Smithsonian Institution, Washington, DC.

[153] Letter from Mitchell E. Giblo to Jacqueline Cochran, September 9, 1964, National Aeronautic Association Archives, Box 139, Folder 6 (Acc. XXXX-0209). Archives Department, National Air and Space Museum, Smithsonian Institution, Washington, DC.

[154] Ibid.

more credit that she had received. She would recommend a team of individuals who were ready to stand behind Joan and offer their support. She wrote:

> *You ask what has been done to have the FAA recognize Joan's longest solo flight in history. Many are concerned that the recognition wasn't immediate, an at least equal to the lesser (in distance) northern hemisphere flight of Jerrie Mock. A race between the two was never intended. Both women deserve great recognition, but for different reasons. I'm anxious now to see the book come out as soon as possible.*
>
> *I'm sure you remember that the FAA presented its award to Betty Miller for her Pacific flight, even though there was no NAA sanction. Mr. Halaby in private has expressed an antipathy to women in flying. So apparently he is averse to giving them recognition when it can be avoided. Betty now has the Harmon Trophy also. Next year, many of us would like to see it go to Joan.*
>
> *One of the most ardent supporters of the world's longest solo flight is Lloyd Spangenberg, editor of* Western Aviation *read by over 35,000. He would like to see FAA recognition for the records Joan's flight set. Some who might well serve on a committee for this essential aviation recognition are Faye Gillis Wells (White House reporter, Washington DC), Grace Page of Daytona Beach, Shirley Marshall of Arizona, Fran Johnson of Las Vegas, Vera Bratz and Lee Winfield of Florida, Lillian Huber of New Orleans, Mary Fields of Oakland, Paul Mantz, and any number of aviation corporations including AC Spark Plug and Al Gossett, Senator Smathers of Florida who holds the FAA purse strings, Ruth Deerman, international president of the 99s, and assuredly, yours truly.*
>
> *Trixie*[155]

---

[155] Letter from Trixie Schubert to Ruth Deerman, November 1, 1964.

Following Joan and Trixie's death, the support would continue. On March 14, 1965, the Ninety-Nines wrote a letter to Colonel Giblo, recommending that Joan be considered for the Harmon Trophy: "We hope the Trustees will agree with us that her accomplishment is indeed worthy of this recognition."[156] The mailing included a detailed overview of Joan's flight experience and career accomplishments. On March 22, 1965, Colonel Giblo responded saying that he would forward the recommendation along to the Harmon Trophies Advisory Committee.

The Harmon Trophies

The Harmon International Aviation Trophies were founded in 1926 by the late pioneer American aviator and balloonist Colonel B. Harmon, with the purpose of promoting greater international understanding and world peace through recognition of outstanding aeronautical achievement, regardless of nationality. Past recipients of this prestigious international honor include Amelia Earhart (1935), Jacqueline Cochran (a multiple-year winner), and Betty Miller (1963). The trophy for female aviators is beautifully elaborate and otherworldly, depicting a winged goddess cradling a falcon with outstretched wings.

On December 14, 1965, Jack Smith flew to the White House (while on active duty in Vietnam) to accept the Harmon International Trophy on Joan's behalf. Two trophies were awarded that year. The Aviator's Trophy went to Max Conrad for achieving a fifty-seven-hour, nonstop, solo flight from Cape Town, South Africa, to St. Petersburg, Florida, establishing a new official world nonstop record for Class-5 lightweight aircraft. The second trophy went to Joan, who became the second general aviation pilot to win the Aviatrix Trophy for accomplishing her 27,750 mile solo flight around the world. As Jack explains in a video interview from December of 2016:

> I had notified the Navy I was going to retire because, when [Joan] finished, we had a great plan to go into business and use her notoriety to . . . we were going to set up a specialized air taxi service

---

[156] McNabb.

for businesses only to have airplanes and pilots ready, when they were ready, so you wouldn't have to fool with the airlines. We had financial backers ready and everything....

At that time I was in the country of Vietnam, and actually the Navy flew me back to Washington to the White House to receive from the vice president her recognition for that. I got a two-week break at Christmas time in the middle of a one-year tour in Vietnam. What they did there was recognize the balloonist of the year worldwide, a military pilot worldwide, a civilian pilot, and a female pilot. She was selected as the Aviatrix of the Year.

At the ceremony, the Ninety-Nines presented a memory book they'd compiled to commemorate Joan's flight, a testament to the immense international goodwill created by it. The book included a letter from the president, as well as a letter in Joan's honor from every prime minister or chief of state from each nation that she landed in during her trip.[157]

Aviation: A Metaphor for Being Bold

According to fellow female pilot and longtime Ninety-Nines member Shirley Thom, who used to attend Ninety-Nines meetings at Trixie's house, being a woman in aviation is fundamentally underscored by a love for learning, adventure, and new experiences. It's about the joy of discovering what lies just beyond the horizon, being open to the gift of learning, and the feeling of being free high above the clouds.

From Amelia Earhart, who became the first woman to fly solo across the Atlantic to Betty Miller, who became the first woman to fly solo across the Pacific; from Mary Petre Bruce and Elly Beinhorn, who turned their idea of using a plane to travel around the world into a reality, to Jacqueline Cochran, the first woman to fly a bomber across the Atlantic; from Ann Holtgren Pellegreno, who completed a commemorative Amelia

---

[157] "Remarks, Vice President Hubert Humphrey, Harmon International Aviation Trophies," Michigan Historical Society, December 14, 1965, http://www2.mnhs.org/library/findaids/00442/pdfa/00442-01768.pdf.

Earhart flight with a crew aboard a Lockheed 10A Electra in 1967, to Linda Finch, who completed a flight commemorating Amelia Earhart on May 28, 1997; and all of the other women who have gone on to complete world solo flights, these women carry the spirit of adventure and bravery forward and exemplify grace in the face of the unknown.

Collectively, these women are not the type who just sit around and wait for things to happen to them; rather, they get out into the world and *make* things happen. They live their lives by engaging with the world rather than being fearful of it. Their courage is not highlighted by an absence of fear, but rather it is based on a commitment to something greater that diminishes those fears. They remind us that there's always a new stone to be unturned, a new surprise waiting just around the bend. It's about appreciating the here and now and soaking up the experience, as much as it is about taking risks at the expense of growth. It's about absorbing the gift of learning, finding out the answers to their questions for themselves, always moving forward and exploring, adapting, progressing.

In the landmark 1926 book *Being and Time*, German philosopher Martin Heidegger explores the concept of what is meant by the phrase "to be" and specifically focuses on the concepts of authenticity, inauthenticity, and fallen-ness. In philosophy, authenticity is defined in terms of the conscious self coming to terms with being in a material world and encountering external forces, pressures, and influences that are very different from, and other than, itself. Authenticity is the degree to which one is true to one's own personality, spirit, or character, despite these pressures. Hence, inauthenticity would be characterized by "falling," or the act of caving in to the external pressures that both distract us from getting to know and staying in touch with our true selves. Falling is characterized by ignoring that inner voice, compromising a piece of your true self, and following the crowd to fit in. To *be* in this world and experience life as it was meant to be lived requires authenticity. Falling is a trap.

Upon graduating from college, one of my dearest professors, Dr. Greg Tropea, gave me a book he wrote entitled *Religion, Ideology, and Heidegger's Concept of Falling*. Originally written as a dissertation for his

Ph.D. program at Syracuse University in 1987, the book deeply explores Heidegger's concepts in *Being and Time* and builds upon the depth of his philosophy. Upon finishing that book, I realized that it was also always meant to be an invitation to consider the importance of how we engage in the world around us. Greg not only inspired his students to dig deep, question everything, think critically, challenge oneself intellectually, and embrace the abstract, but he also reminded us to make time for people, live openly, give generously, and act kindly. His book encapsulated all of this.

Before he passed away in 2010, Greg left behind a personal note that would also double as a call to action. Because I began this book with the sharing of a letter, it seems appropriate here to end with one. He wrote:

I'm swamped with grading, Tiffany, but one of the things you said in your note brought me back to our conversation last September. In September, we talked about whether living a good life was enough, and as you recall, we agreed that ethics alone would not satisfy the spirit. I think that conversation connects to your thoughts about putting the important things on hold to take care of the mundane details first. Each of these thoughts, in its own way, seems to me in danger of leaving the spirit out of the picture, the former by not knowing what it is missing and the latter by deciding for that absence.

The mundane details will always be there. Sure, some greater measure of stability may be achieved, but remember the Buddha's observation that existence has the nature of dukkha, which we can think of as "It's always something." As I see it, putting your best intuitions on hold is like saying you'll listen to the lower harmonics of a piece of music now and the higher harmonics later. In neither case will there be much satisfaction or understanding.

Our entire life is our learning experience. I don't think I'm telling you anything new in saying that the course a life takes depends in large measure upon the soul's attunement. It affects how you perceive and relate to everything. Moreover, as Sartre reminds us, while the activities of the day may be mostly determined by the demands of our professional and personal involvements, HOW we engage in those activities is our choice to make.

So I am not suggesting unrealism or that pitiful new age caricature of wishful thinking. Working constantly for the fullest spectrum of consciousness you can attain will give you a perspective and vision at key decision points that mere strategists will never have. It's not an either-or, it's a both-and. Having your calculations and skills in order, like living the ethical life, is the minimum requirement, not the fulfillment of our life's promise. Lifetimes go by in the blink of an eye and people [can] get lost in the world. I don't want that to happen to you and I know you don't either.

As ever,
Greg

As Greg suggests, *how* we engage in the world is ultimately our own choice to make. I still remember the very first day in 1999 that I sat in Greg's Eastern Philosophy class, sizing him up as strange. With long, gray hair pulled back into a ponytail, sporting a long, dark trench coat, I instantly made the rash assumption that he and I would find little in common. But in no time at all I was completely enamored with his vocabulary, unique insights, inquisitive nature, and deep knowledge.

Much like Greg, rather than conforming to the expectations of their day, Joan and Trixie followed their intuitions and believed in their contributions, which led them to remain true to themselves and achieve great things, each in her own unique and important way. Together, Joan and Trixie were more than pioneering female pilots. Each carried a deep sense of purpose and an unwavering commitment to explore arenas where few women dared. Enlightened and supported by an inner knowledge, each woman made a unique impact and reminds us to always be true to ourselves and never stop learning. This is the essence of what is meant Heidegger's concept of "to be."

As Amelia Earhart famously stated, "Never do things others can do and will do if there are things others cannot do or will not do." So, what are YOU waiting for?

EPILOGUE

When I first took on this story, I was in complete denial about how much of it was actually missing from the original manuscript. Instead of worrying too much about it, though (at the time I wasn't able to even comprehend what was missing since I didn't know the story), I decided to put my head down and get to work because I really wanted to understand it fully. The more I researched, the more I faced unplanned scope creep, but also, the more this research would become a passion project. The best I could tell was that the story I had pieced together from Trixie was extremely compelling, but what I badly needed was context in order to better understand the overall framework.

Meshing the intent of a book written in 1964 with the available breadth of perspective gained more than fifty years later was part art and part science. Sorting through facts and opinions to arrive at answers was a challenge. But, in doing so, certain ideas naturally rose to the surface while others quickly sank to the bottom. Though I had originally put this story aside, waiting on others to lend their knowledge, opinions, memories, and insights, I eventually decided that giving myself permission to move forward with the story I had—Trixie's story—was all I ever really needed. After all, this was a story about Trixie's belief in Joan, their shared belief in adventure, and the power of asking questions, exploring, and finding out the answers for one's own self.

Only after finishing my first draft of the book did I have the heart to go back in and honestly add up all of the pieces that were missing. In the original manuscript, there appear to have been sixteen chapters, and nearly half of them are gone:

- Chapters 1-4 do not exist; I can only deduce that they had to do with Joan's life leading up to the world flight. In their place is a timeline of correspondence between the NAA and Joan.
- Chapter 5 begins with Joan's takeoff from Long Beach and ends in Florida, but Chapters 6 and 7 about her time in Florida through Puerto Rico and on to Brazil are missing.
- Two pages are missing from the original Chapter 8 about Joan's time traveling between Natal, Brazil, and Dakar, Africa.
- Four single-spaced, typed pages about time spent in Lae and Saipan are missing from Chapter 13.
- Chapter 15, which covers her final leg home from Hawaii to Oakland, is also missing.

Whether it turns out that key pages from Trixie's original manuscript were inadvertently or deliberately missing; whether Amelia Earhart's plane is recovered from the Pacific or more concrete evidence is unveiled about her disappearance; whether Joan's two plane crashes were purely accidental or the result of something more nefarious; whether Trixie's book was as yet unpublished because of some normal roadblocks or because someone didn't want Joan's story released; and whether the NAA sanctioning process was fair, unfair, or just one giant misunderstanding, none of those answers are why I have written this book. Rather, my purpose is to help you think more about the stories in your own life—the ones that quietly tug at you day after day, year after year, beckoning you notice them, take a closer look, jump in and explore. Too often, we accept the stories we've been told at face value, without ever really scratching beneath the surface.

What I came to understand throughout this process is it does no good to be a passive bystander. Learning most certainly is not linear. Occam's razor, the philosophical line of reasoning that says the simplest answer is often the correct one, is a useful tool . . . but life isn't always that simple. Asking one question rarely provides you the full, right answer, unless it's the one you wanted to hear (though it does get you a little closer to the truth). There's also no substitute for being curious. Curiosity will propel you in ways that pragmatic questioning alone will never accomplish.

So what did the experience give me that a simple reprinting of the original book would not have accomplished? In short: real depth, a multidimensional perspective, an active role in the story, and the feeling of history coming alive. While I may not have any of the actual answers, I walked away with a much greater appreciation of history, a vivid picture of what Joan and Trixie both accomplished in their time, and a renewed understanding of how past choices can tangibly shape future generations. I learned—even if unwittingly—that the more pieces of the puzzle you can locate, the more the final picture comes into focus.

Underscoring all of this were my continuous and illuminating conversations with others whom I came to know better through this process. The discussions and exchanges I participated in infused and improved this story in ways that facts and research alone never could. It also gave me the wonderful opportunity to connect with people from different walks of life and backgrounds whom I otherwise would never have had the pleasure of knowing. Above all, I received the gift of learning, the pull of adventuring through a project, the joy of wondering what might be waiting for me to find just around the bend. In the end, I think, it felt a bit like the lure of aviation. And I'm guessing that is precisely what Joan and Trixie would have wanted.

ACKNOWLEDGEMENTS

First and foremost, thank you to Heidi Syslo for being simply indispensable throughout this entire process; to Kevin Page for his constant and most appreciated guidance; to Jessica Santina for sticking with me when all of this started off as a crazy and unorganized idea; and to Matt Brown for being thoroughly supportive in all aspects from start to finish. If not for these four, this book simply would never have come together in this way.

Thanks also to Patrice Vacek and Norman Schubert for digging through old files, photos, and memories; to fellow Ninety-Nines members Claire Petrie, Lynn Meadows, and Shirley Thom for their fundamentally important conversations; to Jack Smith for being so gracious well into his nineties; and to Vic Campbell for introducing me to Jack; to Ric Lambart for providing me with clarity, feedback, direction, and support; to Mike Campbell for sharing his insights and opinions, and for suggesting that I think twice about *not* including footnotes; to Claude Meunier for educating me about world solo flights, including his own in 1996; to Joan Darlene for her pivotal guidance at a critical turning point; to Taylor Phillips for his commitment to helping me get important information ironed out; and to Bret Simmons for being the one who encouraged me to start a blog in the first place.

LIST OF APPENDICES

A. Letter from Trixie to Ruth Deerman, November 1, 1964
B. Letter from Hillel Black to Trixie, February 17, 1965
C. Trixie's timeline of correspondence between Joan and the NAA
D. Letter from Roy A. Wykoff to *General Aviation News*, July 27, 1963
E. Witness Account of Airplane Crash Involving Trixie-Ann Schubert and Joan Smith, February 21, 1965

APPENDIX A
LETTER FROM TRIXIE TO RUTH DEERMAN, NOVEMBER 1, 1964

>1967 Micheltorena
>Los Angeles, Calif. 90039
>
>November 1, 1964

Ruth Deerman
Internat'l. Pres. 99s
405 Camino Real
El Paso, Texas

Dear Ruth:

Thank you for your inquiry regarding the book on Joan's flight. I completed it last night, but it required some literal night and day writing. I have a publisher who is most interested, and Lowell Thomas has agreed to write a foreword for the book. It's called:

>WORLD FLIGHT: JOAN MERRIAM SMITH
>Amelia Earhart Route
>
>as told to Beatrice-Ann Schubert

Publishers have been known to change titles so we only assume this will be the published title.

You ask what has been done to have the FAA recognize Joan's longest solo flight in history. Many are concerned that the recognition wasn't immediate and at least equal to the lesser (in distance) northern hemisphere flight of Jerrie Mock. A race between the two never was intended. Both women deserve great recognition, but for DIFFERENT accomplishments.

I'm anxious now to see the book come out soon as possible.

I'm sure you remember that the FAA presented its award to Betty Miller for her Pacific Flight, even though there was no NAA sanction. Mr. Halaby in private has expressed an antipathy to women in flying. So apparently he is adverse to giving them recognition when it can be avoided. Betty now has the Harmon Trophy also. Next year many of us would like to see it go to Joan.

One of the most ardent supporters of the world's longest solo flight is Lloyd Spangenberg, editor of Western Aviation read by over 35,000. He would like to see FAA recognition for the records Joan's flight set. (see accompanying sheet).

Some who might well serve on a committee for this essential aviation recognition are Faye Gillis Wells (White House reporter, Wash. D.C.), Grace Page of Daytona Beach, Shirley Marshall of Ariz., Fran Johnson of Las Vegas, Vera Bratz and Lee Winfield of Florida, Lillian Huber of New Orleans, Mary Fields of Oakland, Paul Mantz, any number of aviation corporations including AC Spark Plug and Al Gossett, Senator Smathers of Florida who holds the FAA purse strings, Ruth Deerman, international president of the 99s, and assuredly,

>Yours Truly,

Joan Merriam Smith Records

First SOLO pilot, man or woman, to circumnavigate the globe at the equator, according to records held at the Smithsonian Institution in Washington D.C., Douglas Aircraft World Flight records, the 1965 World Almanac, and the National Aeronautics Association.

1964   LONGEST SOLO FLIGHT IN HISTORY, 27,750 miles.

First woman to fly a twin-engine plane around the world.

First woman to solo from Africa to Australia.

First woman to solo from Australia to Guam.

First woman to solo from Lake to Midland.

First woman to solo the Pacific with two engines.

First person to receive an ATR rating at age 23.

APPENDIX B
LETTER FROM HILLEL BLACK TO TRIXIE

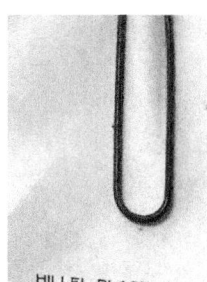

THE CURTIS PUBLISHING COMPANY
641 LEXINGTON AVENUE
NEW YORK, N.Y. 10022
935-3434

HILLEL BLACK
SENIOR EDITOR

THE SATURDAY EVENING POST

February 17, 1965

Mrs. Beatrice-Ann Schubert
1967 Micheltorena
Los Angeles, California 90039

Dear Mrs. Schubert:

Thank you for allowing us to consider the last chapter of your book which describes the death of Joan's airplane. I thought it was well done. However, after some thought and consideration here I have come to the conclusion that it would not work out for us.

But we certainly appreciate your thinking of the Post.

Cordially,

Hillel Black

HB:mc
Encls.

APPENDIX C
TRIXIE'S TIMELINE OF CORRESPONDENCE BETWEEN JOAN AND THE NAA

Joan and NAA

August 4, 1963 wrote NAA for information, asked if any girl had yet filed for the same thing...world flight.
Middle of Sept. asked what was necessary to have world flight sanctioned. (fee was to be paid for NAA rep. at each stop...or pay a NAA rep. to be there.. $50 per day per rep. $1,800 for sanction)
Called again in Oct. "I wanted to go the right route on getting this sanctioned." Chuck Banfe told me it would take about 90 days to get a sanction. I was told to submit a route itinerary.
December 17 I mailed in a complete itinerary. On the 18th my mother fractured her skull from a bus fall and was taken to the hosp.
Jan. 3   Jack left for 7 months overseas and my paternal grandmother died in Miami. (circumstances?)
Jan. 4   Flew to Miami where my mother needed brain surgery to relieve pressure. Got mother out of the hospital on the 8th.

Concerned that there were no sanction papers as I was told I'd have them by return mail so they could be filed out and returned with a $150 sanction fee.
Jan. 7   Jack Riley at Fort Lauderdale. Was looking for a high frequency transceiver. Bob Urico of Sun Aire Electronics at Fort Lauderdale (division of Sun South). He became cagey and asked many leading questions. I said I was planning a flight to Mexico and needed 10 channels. What freqs. do you need? he asked.
The same day I went to Pan Tronics and asked for Ed Neilsson, Not there they said. A radio tech with a foreign accent came out from the back and I asked him about getting some HF equip. "I'm planning a trip out of the country." "Oh, you must be the girl who's going to fly around the world." It was impossible that word could have leaked out on my flight. "Are you the girl from Ohio?" he pursued. (Pan Tronics was installing HF in the Ohio plane.) I played it as if it were the greatest thing since French toast.

I had planned to leave in May at first, from Oakland, as Gerhart did the second time. Reports from meteorologists around the world said the weather was no good that late in some spots.

It was now January and I had nothing but the plane and the promised of $1,000 and some turbo chargers.
I called NAA on Jan 7 at noon and M.J. Randleman answered (secretary of contest board). I had talked with him and written him often before. He told me he had no information regarding another woman's seeking a world flight sanction. I pressed the issued, "If you're working with another woman tell me. Because at this point I haven't the money and don't know whether I should go ahead." (wouldn't you go ahead even without sanction??)
"I'm just getting ready to send out sanction forms," he finally said, "I can say we have received some interest from another girl a year ago."
Mock got NAA sanction. Her sanction forms had not yet been received by NAA. She was asking for sanction from Feb. 15 through May 15. Never had the problem of two flights at the same time come up before. "So we'll have to make a decision which one to sanction. We'll look at them todather."
I said, "The flights are unalike. How can you measure them as alike?"
Randy, "The other flight is only for the Northern Hemisphere. We consider which person is better backed, and better experienced pilot, and with a safer plane...all this we consider. Then the papers and check are sent to the FAI."
"Why don't I get on the airlines and come to Wash D.C. now, fill out the papers and give you $150 check." R., "It's not necessary. Even if you're first in with the papers we'd have to make a decision."

Miller flew without sanction.

Jan. 8 9 a.m. I phoned R. from Miami (my NAA phone call bill was
        now over $300.) Was told, "Her papers came in the morning
        It's over my head now." We can't grant your sanction now.

        The two requests never were considered side by side.

Mock took 30 days around with estimated elapsed time 155 hours.
Joan took   days   170 hours.

Mock has world speed record feminine and world speed record men and w
        (how much better time that Wiley Post??)

Conrad set speed record going 8 days around in an Apached. He flew
    westbound and had some tailwinds. To beat that record I'd have to
    lose about $1,800 and break his time. When would I talk to people
    who had known AE, round up her mechanics etc. And anyway I couldn't
    go west. Earhart didn't.   (prevailing westerlies???)

Mock requested women's speed record only from NAA.

        speed over closed course and around the world
              record (elapsed time) ......
        altitude   ....   NAA sanctions.
        distance (over measured course.....
NAA doesn't have record around the world for distance...only speed.
NAA plane categories, 2300 lbs to 3850 gross.
                      3850 lbs to 6600
                      unlimited above 6600
NAA doesn't sanction FIRSTS. Proof of performance is all that's
necessary. So, I broke the 1924 record of 26,553 miles by four planes
around world.   36 planes have made world flights to date including
military.
        Only 6 of these (including Joan's) met the equatorial
requirements. And only Joan's was solo

APPENDIX D

LETTER FROM ROY A. WYKOFF TO *GENERAL AVIATION NEWS*, JULY 27, 1963

ROY A. WYKOFF, JR.
MUTUAL NEWS-PIX FEATURES
318 E. 9TH STREET
DAVENPORT, IOWA

News Editor
General Aviation News
Los Angeles 54, Calif.

Sirs:

    The undersighned subscriber to the very fine "General Aviation News" noted the article pertaining to the famous aviatrix Amelia Earheart , as per page # 12--in July24, 1963 issue.
    In regards to the famed flyer, the undersighned interviewed Amelia Earheart in October, 1940 on a small island in the Marianos Islands group - Amelia Earheart stated in Oct. 1940 that she exscaped from the area of Garapan, Saipon Island, in July 7, 1937 after a Japanese anti-aircraft burst had disabled the airplane in which she was flyer with Noonan.
    About July 6, 1937 Amelia Earheart was able to secure a native crew of Marshallese natives of Christian up-bringing who were able to hide her in a fishing canoe of wood.
    The Japanese searched the out-rigger canoe but did not look in the fish cargo , and she was able to effect her exscape to a small island about 100 miles distant of Saipan Island--the smaller island was the base in 1937 of a Christian order of missionaries in 1937 and in October, 1940 the undersighned then a flyer with the Flying Tigers of General Claire Chennault American Volunteer Group of the Asiatic wars, was then in the Marianos Islands in Oct. 1940--I met Amelia Earherat at the mission and she was then teaching school to the Marshallese natives of the Christian Mission which was a branch of the Garapon , Mission then located in the Marianos Islands in the area of east Marianos groups of the former German Mandated Islands called Ladrones or Marianos Islands group.
    Amelia Earheart was then dressed in a checkered shirt and wearing Blue dungeree pants of U.S.Navy fatigue type a blue sailor trousers-- the shirt was red and black checker desighn in October of 1940--she appeared well and healthy in October of 1940 ---
    Amelia Earheart was able to ecscape with the aid of an American sailor then marrooned with her on the Island and who had rescued her from the Japanese in July 4, 1937---- when her plane crashed.
    She stated in Oct. 1940 that Mr. Noonan the radio and navigator on her plane had made an erroneous triangulation and flew her plane on radio signals to Saipon and they discovered that the Japs were then firing at the plane with anti-aircraft of the Jap 77 mm guns which hit the engine and the plane crashed in the water about fifty feet off the shore at the Garapan Bay area.

The Earhear Plane sank in fifty feet of water of from the shote-- with all cargo except some papers which Mr. Noonan had in his possession.

The Japs killed Mr. Noonan the latter Mr. Noonan was executed by a Jap firing squad in the courtyard of the Garapaon Mission which was on the N. e. side of Garapon--the Japs then attempted to find Amelia Earheart who was hiding near the Mission in July 5, -6, 1937 but she was hidden by the American and the British women of the Mission--later the Japs captured a British Missionary womane at the Garapon Mission --the woman was from Liverpool, England and she had in her possession the flying coat of Amelia Earheart-

The Japs assumed that the British woman was Amelia Earheart and executed the British Missionary woman in the courtyard of the Mission with Mr. Noonan, according to story of Amelia Earheart who witnessed the executuons from about ½ mile distance with a binocular glass.

The Japs then went from the Mission and several Korean laborers then burred Mr. Fred Noonans body with the British woman in the Garapon Mission at a place about 50 foot north of the main buuling of the Garapon Saipin Island Mission building in July 6, 1937--the burrial site of Fred Noonan is marked by an iron post imbedded in the ground over the bodies of Mr. ~~Noonan~~ and the British woman and Fred Noonan in July 7, 1937---

The above data related by Amelia Earheart in Octoer, 1940 at Marianos Islands.

        ROY A. WYKOFF, JR.
        MUTUAL NEWS-PIX FEATURES
        318 E. 5TH STREET
        DAVENPORT, IOWA

Sincerely and respectfully yours.,

Roy A. Wykoff, Jr.

Formerly Lt. Col. of the A.V.G., Flying Tigers Airforce, of General Claire Chennault, Asiatic Wars--1939 to 1941---

Formerly war correspondent--

27 July 1963

APPENDIX E
WITNESS ACCOUNT OF AIRPLANE CRASH INVOLVING TRIXIE-ANN SCHUBERT AND JOAN SMITH, FEBRUARY 21, 1965

February 21, 1965
3:30 p.m.

Name of witness to airplane crash of Trixie-Ann Schubert and Joan Smith.

Mr. Robert L. Jones (age 38)
Box
Wrightwood, California

Telephone: CH 9-3532
Mr. Jones is the resident manager of the Southern California Bible Conference

Mr. Jones gave the following account:

"I was out on the conference gounds and happened to look up at the sky and observed a jet stream and was following it with my eye when I noticed a small plane. Then I returned to following the jet stream with my eyes. Then I heard the motor of the small plane revve up, so I turned my head to look at it. I noticed that the right wing had folded back against the plane. Then the plane went into a nose dive and crashed."

"It was a clear day. The small plane was flying high above the mountains. I would estimate about 7,000 feet. When I first observed the small plane it was in no trouble at all, flying in a normal manner, headed Southwest. When I continued to follow the jet stream, my attention was turned back to the small plane because of the motor being revved up."

"I called my wife to come and spot the smoke from the crashed plane with me, that we would be certain of the location. Then we called the fire department and the Forestry Service at Big Pine."

"I got in my car and drove immediately to the scene of the crash. There was one man there when I arrived. I do not know his name. I do not believe that he had seen the plane in the sky, only the smoke and fire caused by the crash. Soon the Fire Department arrived and I helped them put out the fire."

Mr. Jones states that the plane was not on fire in the sky. He also stated that he would estimate that from the time he heard the motor of the plane revve up and saw the right wing folded back against the plane until it nose-dived and crashed would be about 10 to 12 seconds.

Mr. Jones stated that he gave this same testimony to:

Mr. William G. Moore
Air Safety Investigator
Civil Aeronautics Board
Bureau of Safety
Los Angeles International Airport
5820 Avion Drive
Los Angeles, California

EXCEPT that it should be SOUTHWEST direction instead of SOUTHEAST

Office: 646-3910
SP 6-0117

Residence phone: 645-2337

- 2 -

Mr. Jones stated that both wings of the plane were at the scene of the crash

The clock on the crashed plane had stopped at 11:45 a.m.----Feb 17, '65
Wednesday

The plane crashed in the San Gabriel Mountains near the Angeles Crest Highway close to Wrightwood, California.

There is a highway below the scene of the crash and a fireroad immediately above the crash. There were two men on the highway just below the crash that were witnesses to the crash. The Sheryiff in Lancaster has their names.

To reach the scene of the crash:
From Wrightwood, drive to Big Pines Ranger Station and on to GRASSY HOLLOWCAMP, where a dirt road takes off to the right. Drive through the campgrounds and a short distnce beyond on the fireroad. The scene of the crash is just below the fireroad here.

            and Leslie
Notes made by Fay Nelson, 3211 Fernwood Ave., LA 39, on Sunday February 21, 1965 after visiting the scene of the crash and the home of Mr. Jones.

ABOUT THE AUTHOR

Tiffany Brown is a Los Angeles native and marketing communications professional with a BA in journalism from California State University, Chico, and an MBA from the University of Nevada, Reno. Her career includes work for global corporations such as CBRE, Trammell Crow Company, and Warner Bros. Studios. Her grandmother, Trixie Ann Schubert, was a multilingual foreign news correspondent on four continents, as well as an avid pilot, book author, newspaper editor, aviation columnist, radio broadcaster, and mother of three. Schubert's experiences and involvement with pilot Joan Merriam Smith inspired Tiffany to write this, her first book.

Tiffany Brown resides in Reno with her family.

Printed in Great Britain
by Amazon

73236980R00129